SPRINGER SERIES IN PHOTONICS 13

Springer

Berlin
Heidelberg
New York
Hong Kong
London
Milan
Paris
Tokyo

Physics and Astronomy

ONLINE LIBRARY

http://www.springer.de

SPRINGER SERIES IN PHOTONICS

Series Editors: T. Kamiya   B. Monemar   H. Venghaus   Y. Yamamoto

The Springer Series in Photonics covers the entire field of photonics, including theory, experiment, and the technology of photonic devices. The books published in this series give a careful survey of the state-of-the-art in photonic science and technology for all the relevant classes of active and passive photonic components and materials. This series will appeal to researchers, engineers, and advanced students.

Series homepage – http://www.springer.de/phys/books/ssp/

Yoshihiro Hamakawa (Ed.)

Thin-Film Solar Cells

Next Generation Photovoltaics
and Its Applications

With 210 Figures

 Springer

Professor Yoshihiro Hamakawa
Department of Photonics
Faculty of Science and Engineering
Ritsumeikan University
1-1-1 Noji-Higashi
Kusatsu Shiga 525-8577
Japan

Series Editors:

Professor Takeshi Kamiya
Ministry of Education, Culture, Sports,
Science and Technology,
National Institution for Academic Degrees,
3-29-1 Otsuka, Bunkyo-ku,
Tokyo 112-0012, Japan

Dr. Herbert Venghaus
Heinrich-Hertz-Institut
für Nachrichtentechnik Berlin GmbH
Einsteinufer 37
10587 Berlin, Germany

Professor Bo Monemar
Department of Physics
and Measurement Technology
Materials Science Division
Linköping University
58183 Linköping, Sweden

Professor Yoshihisa Yamamoto
Stanford University
Edward L. Ginzton Laboratory
Stanford, CA 94305, USA

ISSN 1437-0379
ISBN 3-540-43945-5 Springer-Verlag Berlin Heidelberg New York

Library of Congress Cataloging-in-Publication Data.
Thin-film solar cells: next generation photovoltaics and its applications/Y. Hamakawa (ed.).
p.cm. – (Springer series in photonics; v. 13) ISBN 3-540-43945-5 (alk. paper)
1. Photovoltaic power generation. 2. Thin films. 3. Solar cells.
I. Hamakawa, Yoshihiro, 1932– . II. Series.
TK 2960.T45 2003   621.31'244–dc22   2003057384

Springer-Verlag Berlin Heidelberg New York
a member of BertelsmannSpringer Science+Business Media GmbH

http://www.springer.de

© Springer-Verlag Berlin Heidelberg 2004
Printed in Germany

Data conversion: Christian Grosche, Hamburg
Final processing: LE-TEX, Leipzig
Cover concept: eStudio Calamar Steinen
Cover production: *design & production* GmbH, Heidelberg

Printed on acid-free paper   SPIN: 10885606   57/3141/tr   5 4 3 2 1 0

Preface

The development of clean energy resources as alternatives to oil has become one of the most important challenges for modern science and technology. The obvious motivation for these efforts is to reduce the air pollution resulting from the mass consumption of fossil fuels and to protect the ecological cycles of the biosystems on Earth. Analyses of future energy usage envision that the energy structure in the 21st century will be characterized as a "Best Mix Age" involving different renewable energy forms.

Among the wide variety of renewable energy projects in progress, photovoltaics is the most promising as a future energy technology. It is pollution-free and abundantly available everywhere in the world, even in space, and can also operate with diffuse light. However, a major barrier impeding the development of large-scale bulk power applications of photovoltaic systems is the high price of solar cell modules. Therefore, reduction of the costs of solar cells is of prime importance. To achieve this objective, tremendous R&D efforts have been made over the past two decades in a wide variety of technical fields ranging from solar-cell materials, cell structure, and mass production processes to the photovoltaic systems themselves. As the result, about an order of magnitude cost reduction has been achieved in the past 10 years. According to a recent survey, however, a further cost reduction, to one fifth of the present level, is necessary if photovoltaic energy is to match the present conventional electricity price. Looking back at the past cost reduction measures, one can identify two remaining technologies for further efficiency improvement and cost reduction. The most promising of these is the thin-film solar-cell technology, which has both material saving and efficiency increase whilst at the same time allowing mass-production. The other approach is to exploit the merits of mass production for all photovoltaic system components such as solar cells and modules, inverters etc., also enhancing the market penetration of concepts such as the photovoltaic roof integrated solar house project, building solar tiles, photovoltaic electricity purchasing system, and others that have been initiated in the last few years.

In the past two decades, remarkable advances have been seen in both the physics and the technology of the tetrahedrally bonded amorphous semiconductors. In particular, much attention has been addressed to exploiting vital properties of these materials such as the controllability of their valence electrons, their excellent photoconductivity and significantly higher optical absorption coefficient as compared with crystalline semiconductors. They thus possess enormous potential as opto-electronic materials. In addition to these essential physical properties, they also possess some unique advantages from the technological viewpoint such as the possibility of mass-production through large-area non-epitaxial growth on any substrate material. This corresponds in a very timely fashion to the requirements for developing a low-cost solar cell as a new energy resource. With the aid of the national and/or semi-national projects for renewable energy development, an accelerating expansion of this field has been witnessed in recent years in both basic physics and technology. This new knowledge also opens up some other new application fields such as TFT LCD, laser printers, electro-photoreceptors, three-dimensional integrated devices, and quantum well devices, etc. The remarkable progress in amorphous silicon alloy systems has induced further interest in understanding the basic physics of thin films of other semiconductors such as CIS (copper indium selenide) and its alloys, and has triggered the birth of new kinds of high efficiency solar cell with multi-band-gap stacked (or tandem) type solar cells for the next generation solar photovoltaics.

Recently, there have been a number of books published on the subject of amorphous semiconductors. The principal concern of all these books has been the physics of amorphous semiconductors with emphasis on the random structure solids, and their electronic and optical properties. Although tremendous R&D efforts have been made in the practical application of amorphous semiconductors recently, there have been neither journals nor books published on the subject of material preparation technologies and device applications in relation to thin film solar cells. The purpose of this volume is to summarize the present status of the device physics of thin film solar cells, to describe application systems and to stimulate scientists and engineers working to advance this young, but very promising, clean energy technology.

It is the editor's earnest hope that this book will be helpful not only to all interested researchers and engineers but also to the directors and project supervisors in this new area. Moreover, the contents of the book will lead to further acceleration of advances in this rapidly expanding technological field. In the selection and preparation of the contents for this volume, many people have assisted with support and expert advice. First the authors would like to express their sincere appreciation to the ministry of Education, Special Research Project Office on "Science and Technology of Amorphous and Nanocrystalline Semiconductors", and also the New Sunshine Project Headquarters Office for releasing their supported research aids to the authors. We would like to acknowledge Dr. Claus E. Ascheron, Physics Editor, Springer-Verlag, for his kind help and advice during the editorial stages.

Kusatsu, Shiga, Japan
In a gentle breeze through green leaves,
Early summer 2003   *Yoshihiro Hamakawa*

Contents

List of Contributors

Satyen K. Deb
National Renewable
Energy Laboratory
1617 Cole Boulevard
Golden, CO 80401-3393
USA
Satyen_Deb@nrel.gov

Kimitoshi Fukae
Canon Ecology R&D Center
4-1-1 Kizugawadai
Kizu-cho, Souraku-gun
Kyoto 619-0281
Japan
fukae.kimitosh@canon.co.jp

Yoshihiro Hamakawa
Department of Photonics
Faculty of Science and Engineering
Ritusmeikan University
1-1-1 Noji-higashi, Kusatsu-shi
Shiga, 525-8577
Japan
hamakawa@se.ritsumei.ac.jp

Ryo Hayashi
Canon Ecology R&D Center
1-1, Kizugawadai, 4-chome
Kizu-cho, Souraku-gun
Kyoto, 619-0281
Japan
hayashi.ryo@canon.co.jp

Makato Konagai
Department of Physical Electronics
Tokyo Institute of Technology
0-okayama, Meguro
Tokyo, 152-8552
Japan
konagai@pe.titech.ac.jp

Michio Kondo
Research Initiative
for Thin Film Silicon Solar Cells
National Institute of Advanced
Industrial Science and Technology
Tsukuba Central 2 Tsukuba
Ibaraki 305-8568
Japan
michio.kondo@aist.go.jp

Katsumi Kushiya
Showa Shell Sekiyu K.K
Central R&D Lab.
123-1 Shimo-kawairi, Atsugi
Kanagawa, 243-0206
Japan
kushiyakatsumi
  @showa-shell.co.jp

Toshihisa Masuda
The Energy Conservation Center,
Japan
3-19-9, Hatchobori, Chuo-ku
Tokyo, 104-0032
Japan
t.matsuda@eccj.or.jp

Akihisa Matsuda
Research Initiative
for Thin Film Silicon Solar Cells
National Institute of Advanced
Industrial Science and Technology
Tsukuba Central 2 Tsukuba
Ibaraki 305-8568
Japan
a.matsuda@aist.go.jp

Nobuaki Mori
Institute of Research
and Innovation
1-6-8, Yushima, Bunkyo-ku
Tokyo, 113-0034
Japan
n-mori@iri.or.jp

Shoichi Nakano
Handai Frontier Research Center
2-1, Yamadaoka,
Suita City,
Osaka 565-0871,
Japan
snakano@frc.handai.jp

Tomonori Nishimoto
Canon Ecology R&D Center
1-1, Kizugawadai, 4-chome
Kizu-cho, Souraku-gun
Kyoto 619-0281,
Japan
nishimoto.tomonori@canon.co.jp

Katsuhiko Nomoto
Solar Systems Department Center
Solar Systems Group
Sharp Corporation
282-1 Hajikami, Shinjo-cho
Kitakatsuragi-gun, Nara 639-2198
Japan
nomotok.tis
  @ex.shinjo.sharp.co.jp

Kyosuke Ogawa
Canon Ecology R&D Center
4-1-1 Kizugawadai
Kizu-cho, Souraku-gun
Kyoto 619-0281,
Japan
ogawa.kiyosuke@canon.co.jp

Hiroaki Okamoto
Department of Systems Innovation
Graduate School of
Engineering Science
Osaka University
Toyonaka, Osaka 560-8531,
Japan
okamoto@ee.es.osaka-u.ac.jp

Keishi Saito
Canon Ecology R&D Center
4-1-1, Kizugawadai
Kizu-cho, Souraku-gun
Kyoto 619-0281
Japan
saito.keishi@canon.co.jp

Hans-Werner Schock
Universität Stuttgart
Institut für Physikalische Elektronik
Pfaffenwaldring 47
70569 Stuttgart
Germany
schock@ipe.uni-stuttgart.de

Makoto Tanaka
Energy Devices Department
Materials and Devices
Development Center
Technology R&D Headquarters

SANYO Electric Co., Ltd.
7-3-2, Ibukidani-higashimachi,
Nishi-ku, Kobe City,
Hyougo 651-2242,
Japan
mtanaka@rd.sanyo.co.jp

Yoshihisa Tawada
PV Technology Research
and Development Division
Kaneka Corporation
3-2-4, Nakanoshima, Kita-ku
Osaka 530-8288
Japan
yoshihisa.tawada@kaneka.co.jp

Takashi Tomita
Solar Systems Department Center
Solar Systems Group
Sharp Corporation
282-1 Hajikami, Shinjo-cho
Kitakatsuragi-gun, Nara 639-2198
Japan
tomita.gae
  @ex.shinjo.sharp.co.jp

Hirosato Yagi
Energy Devices Department
Materials and Devices
Development Center
Technology R&D Headquarters
SANYO Electric Co., Ltd.
7-3-2, Ibukidani-higashimachi,
Nishi-ku, Kobe City,
Hyougo 651-2242,
Japan
h_yagi@rd.sanyo.co.jp

H. Yamagishi
Kaneka Corporation
3-2-4, Nakanoshima, Kita-ku
Osaka 530-8288
Japan
hideo.yamagishi@kaneka.co.jp

K. Yamamoto
Kaneka Corporation
3-2-4, Nakanoshima, Kita-ku
Osaka 530-8288
Japan
kenji.yamamoto@kaneka.co.jp

1

Background and Motivation
for Thin-Film Solar-Cell Development

Yoshihiro Hamakawa

Development of clean energy resources as an alternative to fossil fuels has become one of the most important tasks assigned to modern science and technology in the 21st century. As was well recognized after the Kyoto Protocol, the reason for this strong motivation is to stop air pollution resulting from the mass consumption of fossil fuels and to maintain the ecological cycles of the biosystems on the earth. In this chapter, firstly, the influences of the industrial developments of the energy revolutions since James Watt built the steam engine in the 18th century are examined and discussed. The evolution of the main energy resources from coal (solid), oil (liquid) and LNG, LPG (gas) are closely related not only to the economy of mass production, storage, and transportation but also to environmental issues. In the second section, a brief discussion is given on *"the 3E Trilemma"*, which might be the most important issue for civilization in the 21st century.

Among a wide variety of renewable energy projects in progress, photovoltaics (PV) is the most promising one as a future energy technology. However, a large barrier impeding the expansion of the large-scale power-source application of photovoltaic systems has been the high price of the solar-cell module. One of the solutions to achieve a reduction in this cost is the development of the thin-film solar cell, which saves both materials and energy in the production of cells and modules. Another solution is the full use of large-scale activity by mass production as in the semiconductor LSIC (large-scale integrated circuit). To enhance the utilization of large-scale work, some new strategies are needed such as PV roof construction materials or building solar tiles with government subsidies. In the third section of this chapter, a new strategy for the promotion of renewable energy – Fundamental Principle to Promote New Energy Developments and Utilization – and the action planned for PV technology up to the year 2010 are introduced. In the final part of the chapter, a prospect of future industrializations of photovoltaics is discussed, then possible new ways in which PV system developments can contribute to global environmental issues are introduced.

1.1 Development of Modern Civilization via Energy Revolutions

As has been well known, the industrial revolution was initiated by the invention of the steam engine by James Watt in 1765. However, technological maturity from R&D to market penetration takes 15–30 years normally. The steam ships and the railway network powered by the steam locomotives operated by coal fuel were developed in the 19th century. The same is true of the petroleum fuel age, that is, the invention of the internal combustion engine by Etienne Lenoir in 1860 [1] initiated this area. Then, the petroleum age was dominant in the 20th century with gasoline engines for the automobile industry and the aeroplane traffic network. Figure 1.1 shows changes in civilization with the form of energy resource since the industrial revolution initiated by James Watt in 1765. The energy expenditures of the three main fossil fuels are plotted separately up to the year 2000, and their future prospects are also expressed as the dashed lines by following the renewable energy scenario. From the observation of the relation between the industrial developments and their origins of energy resources, coal $\longrightarrow$ oil $\longrightarrow$ gases including recently, LNG, LPG, the important key issues are considered to be two. One is the convenience to respond to mass consumption of energy, that is, mass-production technology, ease of transport, and storage. The driving force for the energy revolution is based upon the so-called *"Great Principle of Economy"*.

For this reason, the form of the energy resources transformed from solid, liquid to gases. Other reasons are governed by the environmental load. As can be seen from Table 1.1 [2], the pollutant emission factors for the electricity generation on the unit of carbon equivalent gram per kWh are 322.8 for coal, 258.5 for oil, and 178.0 for LNG, respectively. According to the recent energy analysis survey [3] the electricity generation factor in the secondary energy is more than 40% in developed countries, and will increase to 50% during the 21st century. Electrical energy is the most convenient energy form for the above-mentioned reasons, in respect of mass production, transport, and distribution for modern civilization.

1.2 3E-Trilemma and New Energy Strategy

In the discussions of the grand design for 21st century civilization much attention is focused on the 3E-Trilemma. That is, for the activation of economical development (E: *Economy*), we need an increase of the energy expense (E: *Energy*). However, this induces environmental issues (E: *Environment*) by more emissions of pollutant gases. On the contrary, if the political option chooses a

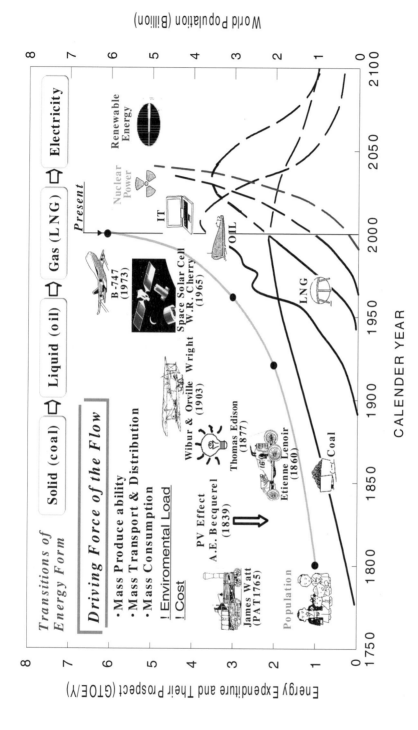

Fig. 1.1. Change in civilization with energy resource

Table 1.1. Pollutant emission factors for electrical generation (g/kWh): The total fuel cycle (TR=trace) [2]

Energy Source	CO_2	NO_2	SO_x
Coal	322.8	1.8	3.400
Oil	258.5	0.88	1.700
Natural gas	178.0	0.9	0.001
Nuclear	7.8	0.003	0.030
Photovoltaic	5.3	0.007	0.020
Biomass	0.0	0.6	0.140
Geothermal	51.5	TR	TR
Wind	6.7	TR	TR
Solar thermal	3.3	TR	TR
Hydropower	5.9	TR	TR

suppression of pollutant-gas emission, it inactivates the economical development. This is the 3E-Trilemma.

Figure 1.2 shows an illustration of the cyclic correlation of the economy, energy, and environment [4]. Here, importance is placed on the change in the circuit from the infrastructure of the fossil fuel's energy supply to that of renewable energy developments and supply. According to the result of world

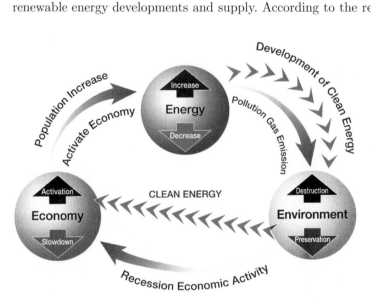

Fig. 1.2. 3-E trilemma, the most important task assigned to 21st century civilization. The only the way to solve this trilemma is developing clean energy technology

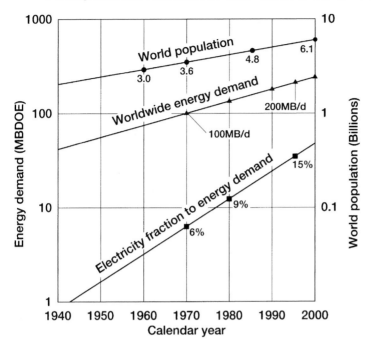

Fig. 1.3. World population growth, worldwide energy demand, and the ratio of electricity to total energy demand (the unit of MBDOE is millions of barrels per day of oil equivalent)

energy trends analysis, the energy consumption per capita in a country is directly proportional to the country's annual income per capital or its GNP (gross national product) [4]. On the other hand, the number of the world's inhabitants is steadily increasing as shown in Fig. 1.3, and reached about 6.1 billion in the year 2000. It can be expected that worldwide energy demand will increase by multiples of the population increase and other factors due to promotion of modernization. This positive increment in energy demand seems unavoidable in the near future even if energy-saving technologies in progress are applied to a moderate degree. For example, the rate of energy consumption per production unit in the heavy industries in well-developed countries might decrease, it will be completely compensated by the rapid increase of energy demand in the newly advancing countries such as China, Malaysia, and Thailand. As can be seen in Fig. 1.4, energy consumption in the East and South Asia area rapidly increases with increasing economic growth, and the percentage of imported fuels will reach almost 70% by 2010 [5].

Considering the two-sided nature of the energy policy, that is, continuous growth of mass consumption of the limited fossil fuels on the one hand, and severe global environmental issues on the other hand, the Agency of Industrial Science and Technology (AIST) in the Ministry of International Trade and

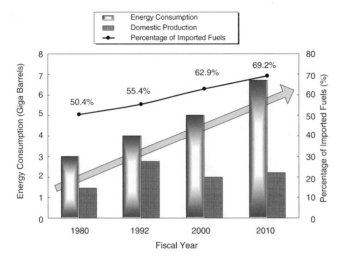

Fig. 1.4. Transition of energy demand in the Asian are (11 countries). The percentages of imported fuels increase along the dashed line

Industry (MITI) in Japan has decided to establish "New Sunshine Program" for the development of clean-energy technology and environmental technology. Figure 1.5 illustrates the comprehensive structure of the new program and its relation to the Sunshine Project, which was formulated in May 1973 prior to the first energy crisis and started in 1974, the Moonlight Project, which was initiated in 1978 for the energy-saving technological development, and also the Environmental Technology Project started in 1989 [6]. Previous programs were a) Renewable Energy Development Technology, b) High Efficiency

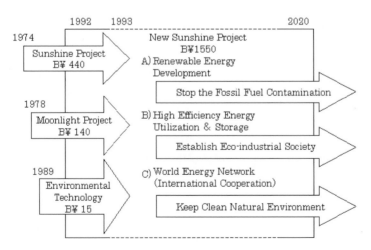

Fig. 1.5. Organizations and objectives of the New Sunshine Project

Utilization of Fossil Fuels and Energy Storage, and c) International Energy Cooperation, the so-called WENET (World Energy Network) utilizing hydrogen fuel-production technology by PV and distributions with a wide-area energy-utilization network system nicknamed "Eco-energy City".

1.3 Key Issues for PV Technology Developments

As can be seen in Fig. 1.2, the most important issue to solve the 3E-Trilemma is development of clean-energy technology, and changes in the circuit from the bold arrow to the dotted arrow. The reason for this strong motivation is to stop air pollution resulting from the mass consumption of fossil fuels and to maintain the ecological cycles of the biosystems on the earth. Views of future energy envision that the energy structure in the 21st century will be characterized as a "Best Mix Age" of different energy forms. Among a wide variety of renewable energy projects in progress, photovoltaics is the most promising one as a future energy technology.

The direct conversion of solar radiation to electricity by photovoltaics has a number of significant advantages as an electricity generator. Solar photovoltaic conversion systems tap an inexhaustible resource that is free of charge and available anywhere in the world. The amount of energy supplied by the sun to the earth is more than five orders of magnitude larger than the world electric power consumption to keep modern civilization going. Roofing-tile photovoltaic generation, for example, saves excess thermal heat and conserves the local heat balance. This means that a considerable reduction of thermal pollution in densely populated city areas can be attained.

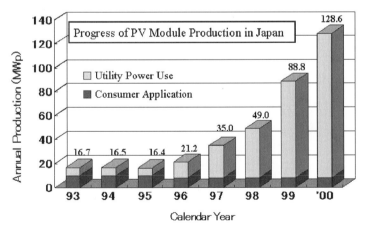

Fig. 1.6. Recent changes in PV solar-module shipment in Japan

Figure 1.6 shows changes in the solar-cell module annual production in Japan since 1993 as surveyed by the Optoelectronic Industry Technology and Development Association (OITDA) [7]. As can be seen from the figure, a remarkable increase of the annual production has been seen since the start of the New Sunshine Project in 1993. In spite of various advantages in photovoltaic power generation, as mentioned above, a large barrier impeding the expansion of large-scale power-source applications was the high price of the solar-cell module, which was more than \$30/Wp (peak watts) in 1974 and \$5.5/Wp even in 1990. That is, the cost of the electrical energy generated by solar cells was still very high as compared with that generated by fossil fuels and nuclear power generation. Therefore, the cost reduction of the solar cell is of prime importance. To break through this economic barrier, tremendous R&D efforts have been made in a wide variety of technical fields, from solar-cell material, device structure, and mass-production processes to photovoltaic systems, over the past 20 years. As a result of a recent fifteen-year R&D effort, about one order of magnitude price decrease has been achieved, and now, the module cost has come down to less than \$4/Wp in a firm bid for a large-scale purchase.

In December 1994, a new initiative of Japanese domestic renewable energy strategy *"Fundamental Principles to Promote New Energy Developments and*

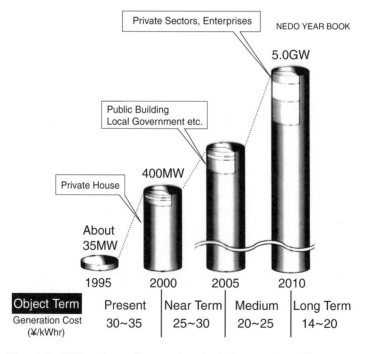

Fig. 1.7. NSS project milestone for the PV-system installation up to 2010

Utilization" was identified by the Cabinet Meeting. Related action planning by law such as tax reduction, government subsidy, etc. were approved by the Congress on April 10, 1997 [8]. The strategies are applied not only to whole ministries and government offices but also local government authorities and private enterprises.

With the new government policy, development and promotion of PV technologies have been suggested as the most promising project. An integrated installations of 400 MWp PV modules by FY2000 and 5.0 GWp by FY2010 for Japanese domestic use are scheduled as a milestone in the program, as shown in Fig. 1.7. A special regulation of tax reduction for investment in renewable energy plant, a government financial support of 1/2 subsidy on the PV system for public facilities so-called *PV field test experiments*, 1/3 subsidy for private solar houses as the *PV house planning* of the field testing, etc., are all in progress.

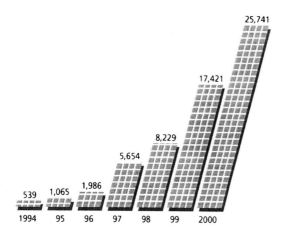

Fig. 1.8. Number of accepted PV house monitor planning by the government subsidy

Figure 1.8 shows the number of accepted PV houses compared to the monitor plan from the government subsidy accepted PV house increase that doubles year by year. On the other hand, in the PV field test experiment a total of 259 sites with 6.84 MW were installed during the 7 years since 1992. A noticeable result in this project is that an accelerated promotion has been done. In fact, the number of installed sites is 73 with 1.94 MW in only one year (1998). As a result of the accelerated promotion strategy, in fact, the sales price for the 3 kW solar photovoltaic system for private houses, for example, decreased very sharply, as shown in Fig. 1.9. As the total system price for a 3 kW solar PV house, 11 MYen in 1993, decreases to 2.4 MYen in 2000, electricity generation cost also falls. This electricity-generation cost will be reduced to a present sales cost of Yen 25.6/kWh in 2005, and will become cheaper than that by hydro-power generation, Yen 18/kWh by the year 2010.

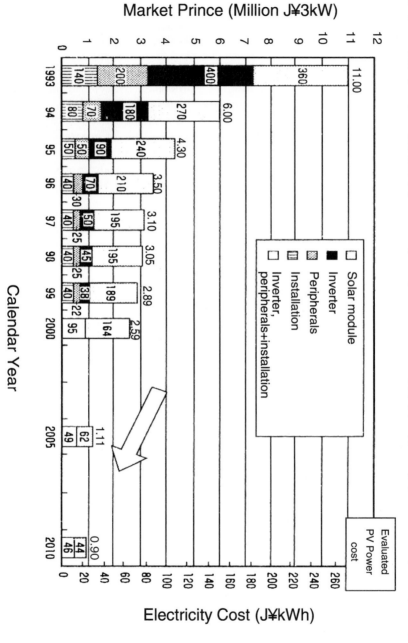

Fig. 1.9. Transition and prospects of the residential PV system cost, a sharp decrease of 3 kW PV roofing facilities and their electricity-generation cost (Yen/kWh) changes in Japan

1.4 Future Prospect and Roadmap for Solar Photovoltaics

As was mentioned in Fig. 1.7, a targeted milestone for volume PV installations by the year 2010 has been set at 5 GWp termed as the *"Fundamental Principles to Promote New Energy Development and Utilization"*. There are many discussions concerning the volume of 5 GW in 2010. For example, if 200,000 solar PV houses were built annually, which corresponds to nearly 10% of the number of private houses built annually, it only takes five years with 5 kW PV/home. Figure 1.10 presents some scenarios of accumulated PV system installations in Japan up to 2030. An estimated installation volume of 65 GW in 2025 in scenario 1 might cover more than 40% of the peak power savings in summer time even taking into account a 12.5% efficiency of a PV system operation. The amount of electrical-power generation of 100 GW in 2030 might easily cover the fossil-fuel electric power-generations in Japan.

Figure 1.11 shows results of a simulation on the future prospect of the primary energy (a) and the electricity generated by various energy resources (b) [2]. As can be seen in the Fig. 1.11a, renewable energy, including PV, will become the major energy resource by about 2050. As for the electric power generation, 2040 will be a critical period for renewable energy in the 21st century.

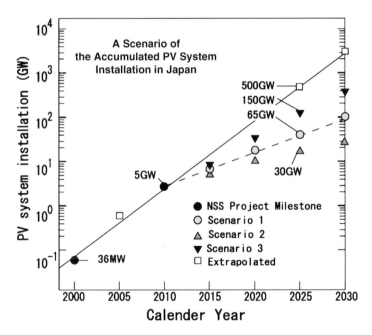

Fig. 1.10. A scenario of the accumulated PV system installation in Japan

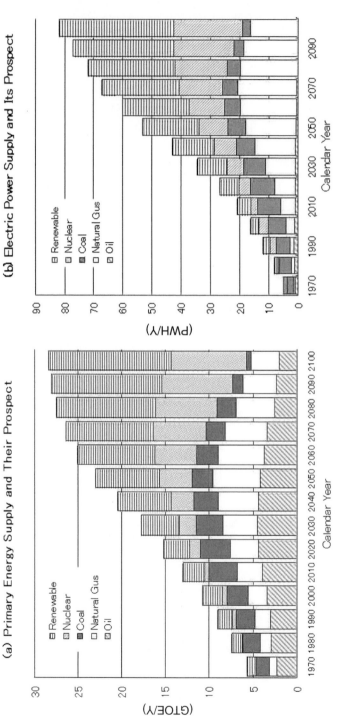

Fig. 1.11. (a) Prospects of the world primary energy demand and (b) that of electric power demand in Japan for th 21st century, calculated for the sustainable development scenario

Table 1.2. Contributions to the global environmental issues by PV

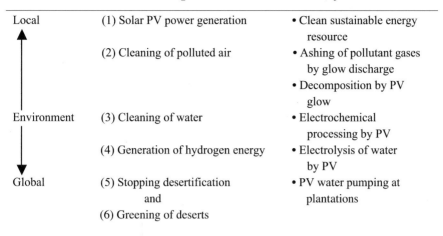

Local	(1) Solar PV power generation	• Clean sustainable energy resource
	(2) Cleaning of polluted air	• Ashing of pollutant gases by glow discharge
		• Decomposition by PV glow
Environment	(3) Cleaning of water	• Electrochemical processing by PV
	(4) Generation of hydrogen energy	• Electrolysis of water by PV
Global	(5) Stopping desertification and	• PV water pumping at plantations
	(6) Greening of deserts	

As described in the previous section, solar photovoltaic power generation is an almost maintenance-free clean-energy technology. Therefore, the penetration of PV technology into utility power generation might be of prime importance for market-size expansion, using a large mass production scale merit of solar cells in the PV system development. With an increase in feasibility of semi-power applications and full use of maintenance-free solar photovoltaics, many methods of antipollution processing can be performed with solar photovoltaic power. For example, ashing of pollutant gas in air by glow discharge decomposition and cleaning of water by electrochemical processing, as shown in Table 1.2 [5]. Mass production of hydrogen energy in the Sahara desert is planned using solar photovoltaics, the so-called VLS-PV (very large scale photovoltaic) [3]. There exists a possibility of "halting desert expansion" and "greening" the desert by photovoltaic water pumping and planting of trees. The "Gobi project" has been organized for this purpose. In 1990, a preliminary study team began a survey of natural conditions as part of a Japan–Mongolia program of cooperation [9].

Results of these considerations indicate that an economically feasible age of photovoltaics will come earlier than expected in the near future, if efforts continue to be made to promote R&D work and market stimulation by means of government support and international cooperation occurs. In any case, the most important emphasis should be placed on stopping the effects of contamination by fossil fuels with a worldwide energy policy and development of this novel, clean-energy technology for the future benefit of all mankind.

The new-energy revolution from the coal age to the oil age was accomplished within only a quarter of a century from 1950 to 1975. The reason why the energy revolution transition time was so short is due to the large-scale merit of oil in terms of mass processing in petrochemical plants. With respect to the scale merit, that of solar cells in a mass-production line might be greater than that of oil at the stage of well-developed PV utilization systems. As has been reported elsewhere, semiconductor devices usually have the highest scale merit, which has been well identified over the past 20 years as initiating a solid-state device revolution from the age of the vacuum tube in the electronics industry. For example, the DRAM (dynamic random access memory) has a 25% in scale merit. This means, if one produces a one order of magnitude increase in production, the cost becomes one quarter. Let us embark on a new kind of energy revolution with clean-energy photovoltaics. Might it be possible to accomplish the clean-energy revolution within the coming 25 years? It is a fanciful dream but let us enjoy our new challenge!

References

1. Agency of Natural Resources and Energy, July 15 (1998), MITI, Energy Flow Diagram of Japan in 1997, Energy Strategy of 21st century.
2. Ito, K. (2000) Prospect of Fossil Fuels in the book of "Energy for 21st Century", 3rd annual Meeting Report for Nuclear Fusion.
3. Kurokawa, K., May (1999) VLS-PV System, IEA-Task IV Report.
4. Hamakawa, Y. (2000): Oyobutsuri 69:8, pp. 993–998 and (1998) ibid 67:9, pp. 1023–1028.
5. Hamakawa, Y. (1999): Proc. WREC Regional meeting, Perth, Australia, p. 37.
6. Miyazawa, K., Katoh, K., and Kawamura, K. (2000): Report of Solar Energy Division.
7. Newsletter 6, OITDA, March 1 (2000) p.1.
8. Meeting, Technology, and Industrial Council, MITI, March 17 (1997).
9. Newspaper Release, Nihon Keizai Shinbun, May 11 (1990).
10. US Solar Energy Industries Association (SEIA), Directory of the US Photovoltaic Industry. 4, (1996).

2

Recent Advances and Future Opportunities for Thin-Film Solar Cells

Satyen K. Deb

2.1 Introduction

The primary objective of worldwide PV solar-cell research and development (R&D) is to reduce the cost of PV modules and systems to a level that will be competitive with conventional ways of generating electric power. To do that, it will be necessary to reduce the cost of PV from $\sim$ \$7/Wp in the year 2000 to about \$1.50/Wp at the system level. Worldwide PV module production in 2000 is estimated to be about 290 MWp, and is growing at the rate of 20–25% per year. The selling prices of PV modules and systems range from \$3–5/Wp and \$6–10/Wp, respectively. PV technology in the marketplace today is dominated by crystalline/polycrystalline silicon. Even with greatly increased production volume and significant reduction of cost, it is doubtful that crystalline or polycrystalline silicon will ever meet the long-term cost goal for utility-scale power generation. If one accepts this rationale, it will be useful to see whether thin-film technologies can realistically meet these goals. There is reason to be very optimistic, particularly in view of the remarkable progress that has been made in recent years in terms of high conversion efficiency, long-term stability, and demonstrated large-scale manufacturing capabilities in several thin-film technologies.

In general, thin-film technologies have the potential for substantial cost advantage versus traditional wafer-based crystalline silicon because of factors such as lower material use, fewer processing steps, and simpler device processing and manufacturing technology for large-area modules and arrays. Because the design and construction of most thin-film PV technologies have many common elements, manufacturing costs are very similar. Therefore, the choice of any given technology is dictated by the highest achievable efficiency, ease of manufacturing, reliability, availability of materials, and environmental sensitivity.

Currently, thin-film technologies based on alloys of amorphous silicon (a-Si:H), cadmium telluride (CdTe), and ternary and multinary copper indium selenide (CIS) are the leading contenders for large-scale production. Figure

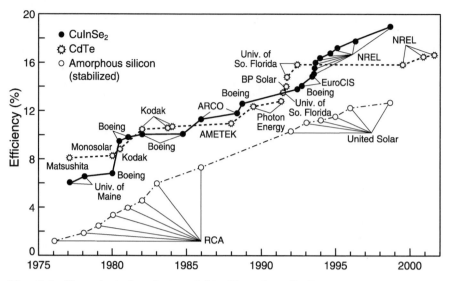

Fig. 2.1. Chronological evolution of thin-film solar-cell efficiencies

2.1 illustrates the chronological evolution of small-area solar-cell efficiencies in these three technologies. In addition to the preceding, thin-film polycrystalline silicon offers an exciting opportunity for low-cost solar cells, and significant progress has been made in recent years in this technology. Currently, the highest-efficiency solar cells have been fabricated using III-V materials and their alloys (thin-film multijunction devices grown epitaxially on single-crystal substrates). However, significant progress has been made on the fabrication of high-efficiency devices on polycrystalline substrate, which raises the hope that high-efficiency thin-film GaAs-based solar cells can be fabricated on low-cost substrates.

There are also exciting new developments in the use of thin-film solar cells based on mesoporous, nanostructure materials such as dye-sensitized (TiO_2) solar cells. The key driver for the future generation of PV is the reduction of cost to about \$1.00/Wp at the system level. In order to achieve this long-term goal, the focus of future PV R&D will be on developing novel thin-film technologies involving new semiconductor materials and devices, as well as pursuing radically new concepts that can significantly enhance the conversion efficiency of a device, and, at the same time, reduce the cost.

2.2 First-Generation Thin-Film Solar Cells

In view of their commercial readiness, the following three groups of solar-cell materials and devices can be considered as the first generation of technologies: (1) hydrogenated amorphous-silicon-based alloy (a-Si:H, a-SiGe:H) p-i-n devices; (2) cadmium telluride/cadmium sulfide (CdTe/CdS); and (3) copper indium gallium selenide/cadmium sulfide (CIGS/CdS)-based heterojunction devices. The technology based on single-junction and multijunction a-Si:H alloys is the leader with a proven manufacturing technology and worldwide production capacity of about 35 MW/y in 2000. Next to a-Si:H, the CdTe-based thin-film solar-cell technology appears to be closest to large-scale commercialization with worldwide production capacity of a little over 1 MW/y. The CIGS-based thin-film technology, although a leader among all three thin-film technologies in terms of cell and module efficiency, is in the early stage of commercialization with current production capacity around 100 kW/y.

2.3 Amorphous Silicon Alloy Solar Cells

Thus far, the thin-film solar cell based on a-Si:H and its alloys as p-i-n junction-type devices are the most mature and commercially available among all of the leading thin-film technologies. As a PV device, a-Si:H alloys offer several advantages: high optical absorption coefficient ($\sim 10^5$ cm^{-1}) with adjustable bandgap (1.1 eV to 2.5 eV) by alloying; ease of fabrication over a large area by a variety of deposition techniques at relatively low temperature ($< 300°C$); multijunction device capability, which ensures optimum utilization of the solar spectrum; and proven cost-effective manufacturing technology. However, the technology also suffers from some major limitations, such as degradation of cell efficiency, usually by 10–20%, due to the Staebler–Wronski effect; degradation of electronic properties at deposition rates greater than a couple of Å/s; and generally low module efficiency. Significant progress has been made in recent years in overcoming some of these limitations, and steady progress has been made in achieving higher efficiency, increased deposition rates, and better photostability.

Unlike other thin-film solar cells, which generally use a p/n junction configuration, a-Si:H-based solar cells require a p/i/n-type of device configuration. This is because a-Si:H and its alloys have intrinsically high defect densities and associated lower minority-carrier lifetime and, therefore, require field assistance for efficient collection of photogenerated carriers. Like other thin-film solar cells, a-Si:H cells are fabricated in two types of configurations – superstrates and substrates, depending on the geometric relationship between the substrate and the rest of the device structure. Currently, the single- or double-junction cells are superstrate type, and triple-junction devices are substrate type. Although major progress has been made in recent years in improving the deposition processes, materials and device quality, and device design and manufacturing processes, the continued improvement of cell efficiency, particularly in single-junction devices, appears to have hit a bottleneck. It is generally recognized that any significant increase in efficiency can be achieved only by a multijunction approach.

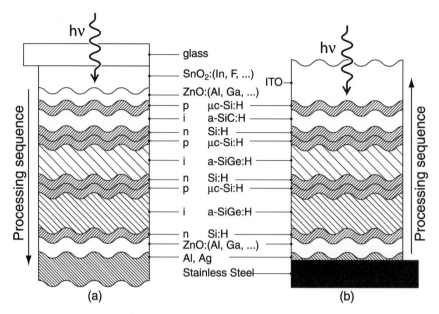

Fig. 2.2. Schematic of a triple-junction a-Si:H solar-cell: (a) substrate and (b) superstrate configuration

2.3.1 Multijunction Cells

Figure 2.2a and 2.2b show schematically a triple-junction device in the substrate and superstrate configurations. The substrate-type devices shown in Fig. 2.2a have been developed principally by United Solar and ECD Corporation and show the highest initial and stable efficiency (13%) reported thus far [1, 2]. The design of a triple-junction cell is based on the well-known principle of spectral splitting such that the top, middle, and bottom cells absorb blue, green, and red photons, respectively. In a typical device, the i-layer of the top cell consists of an a-Si:H with a bandgap of ~ 1.8 eV. The i-layers of the middle and bottom cells consist of a-Si:Ge:H alloys with appropriate composition to adjust the bandgap to ~ 1.6 eV and 1.4 eV, respectively. The thickness of each of the i-layers is adjusted to achieve current matching between individual cells, with the top layer limiting the maximum current. Some of the major innovations that led to the achievement of record efficiency include: (1) use of an optimized hydrogen dilution and other deposition conditions to control the microstructure and improve the material properties of i-layers; the control of microstructure close to the amorphous-to-microcrystalline transition region, which was important in achieving the highest efficiency [3]; (2) incorporation of a graded band structure of i-layers of a-Si:Ge:H layers with the lowest-gap material near the p/i interface to facilitate improved hole transport; (3) use of a high-conductivity, low optical absorption microcrystalline p-layer in the top cell, which contributed to increased V_{oc} (> 1.0 V); (4) development of low-loss tunnel junctions; and (5) use of a textured Ag/ZnO back reflector on the stainless steel substrates, which achieves light trapping for maximum optical absorption.

2.3.2 Single- and Double-Junction Cells

Although triple-junction tandem cells hold the promise for achieving the highest stabilized efficiency, there is a strong argument in favor of both single- and double-junction cell structures because of the inherent simplicity in the fabrication of such devices. Significant progress has been made in recent years in achieving higher stabilized efficiencies in both of these structures. The device structures for both single- and double-junction cells are typically superstrate type. The superstrate is a transparent conducting glass, which serves as the top electrode and an anti-reflection coating. In a single-junction device, a p-i-n structure followed by a ZnO buffer layer and metallic conductor (Ag or Al) is fabricated on the superstrate. A typical two-junction superstrate device has the top and bottom cell consisting of a-Si:H (p-i-n) and a-Si:Ge:H (p-i-n), respectively. Some of the optimization methods used for the triple-junction devices discussed above were also used for the single- and double-junction devices. Large-area single- and multijunction modules are fabricated by following an interconnect design scheme that is applicable to all thin-film cells including a-Si:H and maximizes the output power [4]. Table 2.1 shows recent stabilized efficiencies of some representative a-Si:H-based single- and multijunction devices.

Table 2.1. Stabilized efficiency of a few representative a-Si:H-based cells and modules

Device structure	Stabilized efficiency (%)	Area (cm^2)	Organization
Single-junction a-Si:H	9.3	0.25	United Solar
	9.0	1.0	Tokyo Inst. of Tech.
	8.9	1.0	Sanyo
Double-junction a-Si:Ge:H	12.4	0.25	United Solar
a-Si:H/a-SiGe:H	10.6	1.0	Sanyo
	9.1	842	BP Solar
	9.5	1200	Sanyo
	9.3	5000	Sanyo
Triple-junction	13.0	0.25	United Solar
	10.5	905	United Solar
	12.5 (initial)	922	United Solar

2.3.3 Material and Device Issues

Some of the key research and development issues that affect the current progress and future advances of a-Si:H technology are: growth of high-quality materials at higher deposition rates, optimization of narrow-bandgap alloy materials, and understanding and elimination of photodegradation behavior.

2.3.4 Growth of a-Si:H Alloy

Moderate-quality a-Si:H alloy materials and devices are made by many different techniques including RF, DC, very high frequency RF (VHFRF), and microwave-powered plasma-enhanced chemical vapor deposition (PECVD); electron-cyclotron resonance (ECR) glow discharge; photo CVD; sputtering; and atmospheric plasma CVD. Most of these processes require considerable finetuning of their deposition parameters before high-efficiency devices can be fabricated. Among these processes, only DC- or RF-powered PECVD is currently used for the manufacturing of amorphous-silicon solar cells. One of the major drawbacks of the PECVD process is the very low deposition rate, usually around 1–2 Å/s. However, a higher deposition rate is essential for achieving the throughput needed for the reduction of cost. Extensive effort has been made in the past to increase the deposition rates of the i-layer by using the standard PECVD process, but without any success. However, several groups have shown that by using very high frequency (VHF) PECVD, the deposition rates of a-Si:H can be increased to the 10–15 Å/s range without any degradation of the film properties [5, 6].

Recently, an a-Si:H/a-SiGe:H/a-SiGe:H triple-junction solar cell has been fabricated in which the i-layers were deposited at 10 Å/s using VHF-PECVD (70 MHz). They achieved an initial active-area efficiency of 11% at AM 1.5, which degraded to 9.6% after 700 h of light soaking [7]. VHF PECVD is particularly interesting, because the technique can be easily adapted to the existing PECVD processor.

Achieving even higher deposition rates requires alternative deposition processes; two new techniques, hot-wire CVD (HWCVD) and the atmospheric pressure plasma CVD (APP-CVD), appear to have much promise. The HWCVD technique has been successfully developed in recent years to grow device-quality a-Si:H alloy films at deposition rates up to 167 Å/s with excellent structural and electronic properties such as high photosensitivity $(\sigma_p/\sigma_D = 10^5)$, low saturation defect density $(N_d = 2 \times 10^{16}/cm^3)$, high ambipolar diffusion length ($\sim$ 2000 Å), and low hydrogen content ($C_H < 1\%$) [8]. HWCVD films are normally grown at substrate temperatures > 400°C. Implementing this technique for the fabrication of a p-i-n superstrate cell structure is problematic because of the degradation of the p/i interface at this temperature. However, the n/i interface is less sensitive to high temperatures and, therefore, the n-i-p substrate-type structure is the device of choice. Recently, a hybrid process in which n-i layers are deposited by HWCVD and the p-layer is deposited by PECVD produced devices with initial efficiencies of 9.8%. By using a lower substrate temperature (280–400°C) and an all-HWCVD process, an efficiency of 8.7% was achieved with the i-layers grown at rates of 17–20 Å/s [9]. Contrary to early expectations, the solar cells fabricated by the HWCVD process also show significant photoinduced degradation.

An alternative high-rate deposition technique involving atmospheric-pressure plasma CVD with a rotary electrode has been developed for the fabrication of high-quality a-Si:H films at very high deposition rates (0.3 μm/s to 1.6 μm/s) [10]. A 150-MHz VHF power was used to generate high-density plasma in a gas mixture containing He, H_2, and SiH_4 for this process. The films deposited by this technique show photosensitivity (σ_p/σ_D) of 10^6 at a very high deposition rate of 0.3 μm/s. The process has yet to be applied to the fabrication of a solar cell.

Besides the high-rate deposition processes, other significant advances have been achieved involving the deposition of high-quality a-Si:H film by the use of hydrogen dilution during film growth. An optimized hydrogen dilution leads to higher-quality film at deposition rates up to 15 Å/s [11]. A record 9.3% stabilized efficiency has been achieved in a large-area (5150 cm^2) a-Si:H/a-SiGe:H double-junction cell using optimized H-dilution for the deposition of the absorber films. Hydrogen dilution also leads to an amorphous-to-μ-crystalline transition, and the best solar-cell material is obtained when the film is deposited below this transition region [3].

Futako et al. [12, 13] have developed another novel preparation method called "chemical annealing" for tuning the bandgap of a-Si:H film in the range of 1.5–2.0 eV. The process involves treating the growing film with chemically reactive species such as atomic H or electronically excited Ar to enhance structural relaxation.

2.3.5 Photodegradation of a-Si Solar Cells

The photoinduced creation of metastable defects, the so-called Staebler–Wronski (S–W) effect, in amorphous silicon presents a formidable challenge to the improvement of solar-cell technology. The effect is mainly observed through the degradation of electronic properties due to the formation of deep-level defect states that act as recombination sites. These defects are metastable and can be removed by thermal annealing above 150°C for several hours. This degradation/recovery process can be repeated many times. Although solution of this instability problem remains elusive, a partial solution has been found that involves the use of a thinner a-Si:H intrinsic layer. Thus, by reducing the i-layer thickness, while increasing the optical absorption by effective light trapping, it is possible to ameliorate the stability problem. It is also observed that a multijunction approach, which uses thinner i-layers, gives rise to a higher stabilized efficiency than a single-junction cell. In recent years, significant progress has also been made in improving the material properties by controlling various deposition-process parameters and film microstructures, which has contributed to the improvement of device stability. Most high-efficiency, triple-junction devices show degradation of about 10% as compared to single-junction devices, which used to degrade, in some cases, by as much as 50%. Long-term light soaking of a-Si:H PV modules indicates that stabilization of degradation depends on the operating condition [14].

Extensive studies have been made during the last two decades to understand the mechanism of this metastability and to minimize its impact. An excellent review of this effect in the a-Si:H alloy describes our current understanding of this issue [15].

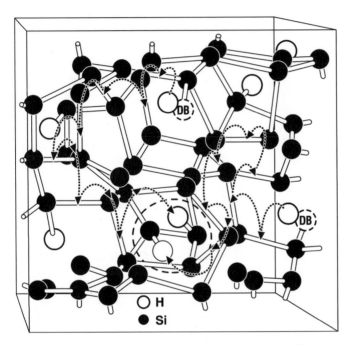

Fig. 2.3. Hydrogen-collision model for photodegradation

In the past two decades, extensive effort in developing models for this effect has led to the emergence of two models, (1) the "charge-transfer model" developed by Adler in 1983 [16] and, (2) the "hydrogen-bond switching" model proposed by Stutzman et al. in 1985 [17]. Recently, Branz [18] suggested the "hydrogen-collision model" shown in Fig. 2.3, which can explain many aspects of the S–W effect. This model assumes that optical excitation creates two mobile hydrogen atoms that collide and form metastable two-H complexes $M(Si-H)_2$ and metastable dangling bonds (DB) according to the equation 2 Si-H → 2 DB + 2 Si-H/DB → 2 DB + M $(Si-H)_2$. Theoretical analyses, using molecular dynamic simulations, have supported the feasibility of formation of this defect complex [19]. The first direct evidence of the light-induced, long-range hydrogen motion at room temperature in a-Si:H is demonstrated by studying the photoinduced diffusion of H from a-Si:H to WO_3 film in a bilayer structure of a-Si:H/a-WO_3 [20].

2.3.6 Technology Development

Well over a decade of technology development for the large-scale commercialization of a-Si:H-based solar cells has given this technology an early start over other competing thin-film devices. In the United States, several companies have been involved in this area. Some of the leading ones are United Solar System Corporation (USSC), Energy Conversion Devices (ECD), BP Solar, and Iowa Thin Films Technologies. BP Solar built a 10-MW a-Si:H/a-SiGe:H tandem-junction cell manufacturing plant as shown in Fig. 2.4. The two-junction device structure consists of a p-i-n-p-i-n substrate-type device structure on glass. The dc-glow-discharge PECVD process is used to deposit the a-Si:H and a-SiGe:H thin films. The final products consist of 5-W to $\sim$ 50-W modules [21]. USSC and ECD, on the other hand, manufacture a superstrate-type, triple-junction device in a roll-to-roll process in which half-mile-long, 14"-wide, and 5-mill-thick stainless steel is continuously coated to fabricate the complex device structure [22]. The coated web is then cut into 9.4" $\times$ 14" pieces and processed to make a variety of flexible products for a wide range of applications. The standard product ranges from 3 W to 64 W, with a stabilized aperture area efficiency of 7.5 to 7.8%. USSC has also

Fig. 2.4. BP Solar 10-MW manufacturing plant

introduced PV shingle roofing products, which resemble conventional roof
shingles.

2.3.7 Research Issues in a-Si:H-Based Materials and Devices

The key research issues that affect the future of the a-Si:H technology have
been reviewed at a recent workshop [23] and broadly classified into two areas:
materials and devices. The issues related to materials involve novel growth
methods, growth reactions and plasma chemistry, role of hydrogen, high-
and low-bandgap alloy materials, and structural order such as amorphous
$\rightarrow$ nanocrystalline $\rightarrow$ microcrystalline transitions. The device issues are: approaches to increasing throughput and reducing cost, long-term stability, role
of surface and interfaces, fabrication of large-area high-efficiency devices, and
device modeling.

2.4 CdTe-Based Thin-Film Solar Cells

Next to amorphous silicon, CdTe-based thin-film solar cells are closest to
commercial readiness. The CdTe/CdS thin-film solar cell is normally a heterojunction device, in which CdTe is the p-type absorbing layer. The optimum
bandgap (1.44 eV), high absorption coefficient typical of a direct bandgap
semiconductor and other physical properties make it an ideal PV material
with a theoretical efficiency around 30%. Analogous to the Cu_2S/CdS solar
cell, the first heterojunction solar cell involving $Cu_2Te/CdTe$ was fabricated
by Cusano in 1963 [24]. A number of device structures involving a p-n homojunction, semiconductor-insulator-semiconductor junction, and a variety of
heterojunctions have been extensively studied in the past three decades. The
status of the technology up to 1983 has been reviewed by Fahrenbruch and
Bube [25]. CdS/CdTe has emerged as the most promising device structure, in
which CdTe is the p-type absorber layer. Efficiencies in the range of 15–16%
have been achieved in recent years on relatively small-area devices [26, 27].

One of the major advantages of CdTe/CdS thin-film solar cells is the low-
cost fabrication option. A number of relatively simple, low-cost methods have
been used to fabricate solar cells with efficiencies in the range of 10–16%. Some
of the low-cost deposition methods for CdTe that show promise include: (1)
closed-space sublimation (CSS); (2) spray deposition (SD); (3) screen printing
(SP); and (4) electrodeposition (ED). All of these techniques are being pursued
for large-scale manufacturing by several companies.

2.4.1 Device Fabrication Process

Most of the high-efficiency CdTe-based heterojunction devices use a superstrate configuration in which CdTe is deposited on the CdS window layer.

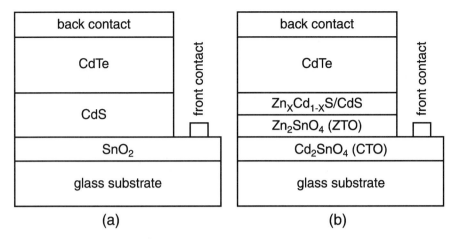

Fig. 2.5. A typical CdTe device structure: (a) conventional, (b) modified version

A typical device structure, which is routinely used for high-efficiency device fabrication, consists of glass/SnO$_2$/CdS/CdTe/ back contact and is shown in Fig. 2.5a. The following sequence of processing steps is generally used for device fabrication. A transparent conducting glass (soda lime or borosilicate glass) coated with doped SnO$_2$ or indium tin oxide (ITO) films is used as a substrate. A thin film (0.05–0.1 μm) of CdS is then deposited by one of a variety of techniques. Both CSS and chemical-bath deposition (CBD) techniques are generally used for the fabrication of high-efficiency devices. The CdS film is heat treated in a reducing atmosphere (H$_2$ or HCl) or in the presence of CdCl$_2$ in the temperature range of 400–500°C.

The p-type CdTe thin film (1.5–3.0 μm) is deposited on the CdS layer by a variety of methods such as CSS, ED, MOCVD, SP, PVD, and sputtering. All of these techniques have led to cell efficiencies ranging from 10–16%. However, techniques such as CSS, ED, and spray deposition are widely used for large-scale manufacturing. Regardless of the deposition method used, the CdTe requires a post-deposition heat treatment with CdCl$_2$ or other Cl-containing compounds to improve its microstructural and electrical properties.

As shown in Fig. 2.6, the CdCl$_2$ treatment leads to considerable grain growth in both CdS and CdTe, which is necessary for high-performance devices. Forming a low-resistance contact to the high-resistivity p-type CdTe film presents a challenge. A variety of approaches have been taken and a number of contacting materials, such as Cu, Au, Cu/Au, and Ni have been used thus far. Some of the limitations on the performance and reproducibility of the conventional CdS/CdTe device are low sheet resistance (10 $\Omega/\square$) and poor light transmittance ($\sim$ 80 %) of SnO$_2$ (TCO). The 2.4 eV bandgap of CdS gives poor blue response, performance reproducibility, and film adhesion during CdCl$_2$ treatment. A modified device structure shown in Fig. 2.5b removes or significantly minimizes most of the limitations discussed above [28]. In this

Fig. 2.6. Morphology of CdTe film before and after CdCl$_2$ treatment

structure, SnO$_2$ is replaced by a Cd$_2$SnO$_4$ (CTO) film with superior electrical and optical properties, and a buffer layer of high-resistivity Zn2SnO$_4$ (ZTO) is sputter deposited on a CTO layer. During the CdCl$_2$ heat treatment of the CdTe film, the ZTO undergoes structural changes (amorphous to crystalline), and, at the same time, interdiffusion of Zn and Cd at the interface between ZTO and CdS, which forms a Zn$_x$Cd$_{1-x}$S layer, occurs, thereby improving the device performance.

2.4.2 Cell and Module Efficiency

Until recently, the highest conversion efficiency of 15.8% was reported by a group at the University of South Florida. More recently, a world record of 16% efficiency at AM1.5 was reported in a CdS/CdTe solar cell by Matsushita Battery Industrial Company of Japan by using ultrathin CdS (0.05 μm) and thin CdTe (3.5 μm) film deposited by the MOCVD and CSS techniques, respectively. The modified device structure, which has the potential for significant enhancement of device performance and reproducibility, has shown 15.8% efficiency. It is interesting to note that a variety of deposition methods have been used to achieve efficiencies ranging from 12–16%, and this is shown in Table 2.2. A modest improvement in cell parameters can easily lead to a cell efficiency of 18.5%.

A number of industrial groups throughout the world are currently engaged in the development of CdTe-based thin-film solar-cell modules. The current status of some representative cells and modules fabricated by different processes is given in Table 2.2.

2.4.3 Technology Development

A number of companies throughout the world are currently involved in the technology development of large-scale manufacturing of CdTe-based devices.

Table 2.2. Efficiencies of CdTe/CdS solar cells and modules

CdTe deposition process	Efficiency/area	Organization
CSS	15.8%/1.05 cm^2	USF
	9.1%/6728 cm^2/61.3 W	Solar Cells, Inc.
CSS	16.0%/1.0 cm^2	Matsushita
Electrodeposition	14.2%/0.02 cm^2	BP Solar
	8.4%/4540 cm^2/28.2 W	BP Solar
Atomic layer epitaxy	14.0%/0.12 cm^2	Microchemistry
Screen printing	12.8%/0.78 cm^2	Matsushita
	8.7%/1200 cm^2/10.0 W	Matsushita
CSS	15.8%	NREL

Some of the key players are: (I) Solar Cells, Inc. (currently First Solar, LCC), ANTEC Solar, BP Solar, and Matsushita Battery Co. First Solar, LLC, is currently working on a light throughput (2–9 m^2/min 100-MW/y production facility [29] with an objective of achieving 10% module efficiency. The schematic of their module structure and the processing sequence is shown in Fig. 2.7. BP Solar has fabricated 38.2-W modules with an efficiency of 8.4% (4540 cm^2 area), and is currently setting up a manufacturing facility in the United States. ANTEC Solar GmbH is in the process of setting up a 10-MWp/y production facility [30].

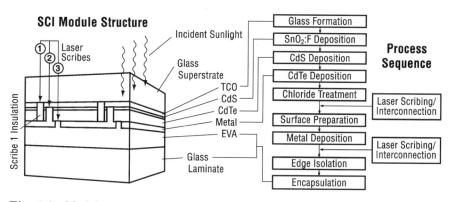

Fig. 2.7. Module structure and processing sequence of CdTe solar cells by Solar Cells, Inc.

2.4.4 Key Issues for Future R&D

Some of the key issues are: (1) basic understanding of the material properties, particularly with regard to the role of structural defects, dopants, and impurities, grain growth and grain boundaries, and different processing parameters; (2) surface and interface phenomena affecting the junction properties, interfacial diffusion, and grain-boundary passivation; (3) optimization of front and back contacts with respect to contact resistance and stability of contacting materials (establishing ohmic contact with p-type CdTe has been a challenging problem); and (4) optimizing online manufacturing processes to close the gap between small- and large-area devices, an issue that is common to all thin-film solar cells. These and other issues have been addressed in a recent workshop [31].

2.5 CuInSe$_2$(CIS)-Based Thin-Film Solar Cells

After almost 25 years of extensive work on CdS/Cu$_2$S thin-film solar cells, this was abandoned in the early 1980s because of insurmountable problems associated with the stability of Cu$_2$S material. The attention was quickly focused on CIS as a replacement for Cu$_2$S because of two reasons: (1) demonstration of a fairly high efficiency $\sim$ 12%) in a single-crystal CIS-based device at Bell Laboratories [32] and (2) fabrication of thin-film devices with reasonable efficiency (6.6%) [33]. Research funded by the U.S. Department of Energy (DOE) quickly led to the achievement of 12% efficiency in CuInSe$_2$/Cd(Zn)S thin-film solar cells by the Boeing Corporation [34]. It was also shown that, unlike Cu$_2$S/CdS, this device had the potential for long-term stability. This led to a worldwide R&D effort, which resulted in major innovations in the materials and device-processing technologies. These innovations included alloying of CIS with Ga to engineer the bandgap for optimum solar conversion, introduction of a back reflector for electrons, use of a gradual bandgap for better charge collection, replacement of the bulk of the CdS film by a ZnO bilayer, a CBD process for CdS deposition, and better understanding and control of the optoelectronic, structural, and defect properties of bulk materials and interfaces. This culminated in the achievement of a record efficiency of 18.8% at NREL in recent years, which puts this technology at the forefront of high-performance thin-film solar cells [35].

2.5.1 Device Fabrication

The device structure that led to the achievement of the highest efficiency (18.8%) consists of glass/Mo/CIGS/CdS/ZnO/metal grid/AR coating and is shown in Fig. 2.8a. It is a substrate-type that is coated with $\sim$ 1 μm layer of Mo deposited by RF sputtering. The choice of substrate, particularly the use of SLG, is important because of the beneficial effect of Na outdiffusion from

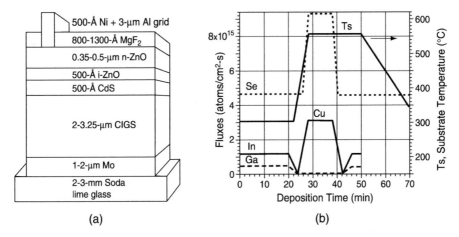

(a)   (b)

Fig. 2.8. 18.8%-efficient CIGS/CdS/ZnO solar-cell: (a) device structure, (b) elemental fluxes and substrate temperature versus deposition time

the substrate into the absorber layer. However, an Mo-coated stainless steel substrate has also been used to fabricate high-efficiency (17.4%) devices. The CIGS absorber layer is deposited on the Mo layer by a three-stage physical vapor deposition method (PVD), which follows a protocol for maintaining the elemental fluxes and substrate temperature profile as a function of deposition time, as shown in Fig. 2.8b [36]. It is a crucial processing step in the sense that it maintains a Cu-poor stoichiometry, overall $Ga/(Ga + In)$ ratio to ~ 0.28, graded profile of Ga, and a film morphology that has the smooth surface and large-grain structure (Fig. 2.9) that are necessary for high-efficiency devices. A thin (~ 500 Å) CdS buffer layer is deposited on the CIGS film by a chemical-

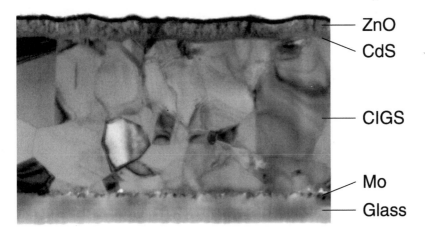

Fig. 2.9. Morphology of CIGS and CdS film

bath deposition (CBD) method. This is followed by the deposition of a high-resistivity/low-resistivity bilayer of ZnO by RF sputtering. Typically, the high-resistivity ZnO layer (0.5 μm) is deposited on the CdS layer from a pure ZnO target and the low-resistivity ZnO layer (0.35 μm) from an Al-doped ZnO target. The device is completed by depositing a Ni/Al collector grid with 5% obscuration and a 0.1-μm-thick MgF_2 antireflection coating. The same device without the CdS buffer layer produces a 15%-efficient device. The performance characteristics of several high-efficiency devices fabricated at NREL are shown in Table 2.3.

Table 2.3. High-efficiency, CIGS-based thin-film solar cells fabricated at NREL

Device structure	Eff. (%)	Cell area (cm^2)
ZnO/CdS/CIGS/Mo	18.8	0.449
ZnO/CdS/CIGS/Mo	17.7	0.414
ZnO/CIGS/Mo	15.0	0.462
ZnO/CdS/CIGS/Mo (Mo-coated stainless steel)	17.4	0.414
ZnO/CdS/CIGS/Mo	13.7	43.89

Very high efficiency CIGS solar cells have been developed in several laboratories in Europe and Japan. In spite of rather complex processing steps involved in the fabrication of the CIGS absorbing layer, the material itself is very robust and defect tolerant. Like CdTe thin films, CIGS can also be fabricated by a variety of processes such as sputtering, spray pyrolysis, electrodeposition, and molecular beam spectroscopy, although PVD appears to be the preferred method. Over the years, several variations of the PVD have been developed for high-efficiency device fabrication.

2.5.2 Technology Development

Excellent progress has been made in recent years on the fabrication of large-area CIGS cells and modules by a number of groups worldwide. Some of these corporations include Siemens Solar Industries (SSI), DayStar Technologies, Energy Photovoltaics (EPV), Global Solar Energy (GSE), and International Solar Energy Technology (ISET) in the United States.; Nordic Solar Energy in Sweden; Matsushita Electric and Showa Shell in Japan; and others. Some representative efficiencies of large-area devices and modules are given in Table 2.4. SSI, the first company in the world to produce CIS modules, achieved a world-record efficiency of 12.1% in a 3651-cm^2 module with output power of 44.3 W. In addition to higher efficiencies, one of the major advantages of CIS-based solar-cell technology is its demonstrated stability over an extended

Table 2.4. Some high-efficiency large-area CIS and CIGS thin-film modules

Absorber film	Area (cm^2)	Efficiency (%)	Power (W)	Company
CIGS	3651	12.1	44.3	Siemens Solar
CIS	3859	10.2	39.3	Siemens Solar
CIGS	717	11.45	-	ZSW
CIGS	51.7	14.2	-	Showa Shell

period of time. Several thin-film solar-cell modules have been tested outdoors at NREL over a period of almost 10 years without any significant degradation.

2.5.3 Key Research Issues

Some of the key research issues are: (1) development of an integrated predictive understanding of CIGS(S) materials and devices, (2) development of novel deposition techniques and characterization of the mechanisms of growth in existing and novel processes, (3) novel materials, especially with wide energy gaps (> 1.7 eV) other than Cu-based chalcopyrites, (4) development of real-time material characterization for process control, and (5) alternative front- and rear-contact materials [37].

2.6 Next Generation of Thin-Film Solar Cells

The next generation of thin-film solar cells covers a broad spectrum of materials, devices, and novel concepts that are in the exploratory state of research and development and have the potential to replace the current technologies. Many of these technologies will be either an extension of current materials and devices or will involve novel concepts that hitherto have been unexplored for solar-cell applications [38–40]. Some of the areas that will be briefly reviewed here involve: silicon-based thin films, thin films of conventional III-V and II-VI semiconductors, dye-sensitized TiO$_2$ solar cells, organic solar cells, and some novel concepts that are in a rudimentary state of development.

2.6.1 Thin-Film Silicon Solar Cells

Thin-film multicrystalline silicon (mc-Si) offers an exciting opportunity for the development of efficient, low-cost, and stable solar cells by virtue of the many positive attributes of silicon as a PV material. However, the primary drawback is the inefficient absorption of solar radiation due to the indirect bandgap of silicon. This disadvantage can, in principle, be overcome by suitable light-trapping schemes. Thin-film silicon has some advantages as compared to its crystalline analog because of the less-stringent requirement on the material quality. Theoretical calculations show that it is possible to achieve an efficiency of around 17% in a 2-μm-thick Si film of appropriate grain size (10μm) and low dislocation density ($< 10^6/cm^2$) [41]. The concept of thin films is not new. Several approaches were developed in the early 1980s for fabricating thin-film silicon solar cells on low-cost substrates, and efficiencies greater than 10% were demonstrated. However, with the emergence of a-Si:H and other promising thin films of compound semiconductors, the field was prematurely abandoned. In recent years, there has been a resurgence of interest in this area since the achievement of 9.3–9.8% efficiencies in polycrystalline Si-film of 8.8 μm and 1.8 μm thickness, respectively, by the Kaneka Group in the so-called 'STAR' (surface, texture-enhanced absorption with a back reflector) cell shown in Fig. 2.10. To design and fabricate efficient, low-cost, thin-film mc-Si solar cells, some major technological challenges need to be addressed: high-throughput growth of thin-film silicon on low-cost substrate and developing methods for growing films with grain size larger than the film thicknesses; and effective light trapping to enhance optical path length. Some of the approaches that are currently being pursued include: use of reusable single-crystal Si substrates by using lift-off and smart-cut techniques; use of multicrystalline-Si wafers; and low-cost ceramic, graphite, glass, or metallic substrates. A variety of techniques similar to those used for a-Si:H, including CVD, PECVD, sputtering, and vapor transport, can be used for the deposition of Si thin film. In general, most of these techniques lead to the growth of films with small grain size (< 1 μm), and subsequent processing is needed to increase the grain size. Some of the processing techniques used are:

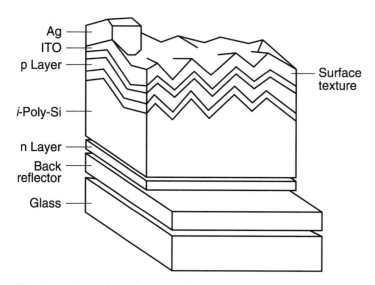

Fig. 2.10. Thin-film silicon 'STAR' cell by Kaneka group

thermal annealing, metal-induced crystallization, zone melting, laser-induced recrystallization, and optical processing. Details of these fabrication and design processes and the current status of the technology are given in a recent review [42].

2.6.2 Microcrystalline-Silicon Thin Film (μc-Si)

Hydrogenated microcrystalline silicon shows very efficient n- and p-type doping properties and, because of this, is widely used as ohmic contact layers in (p-i-n) a-Si:H solar cells. The mc-Si:H grown by VHF-GD deposition shows enhanced optical absorption due to increased surface roughness and preferential columnar-growth morphology. These properties, along with the ability to achieve midgap character by controlled oxygen doping, make it attractive for use as an active absorber layer for solar cells. Recently, the pioneering work at the Institute of Microtechnology (IMB) at Neuchatel has led to the fabrication of an entirely microcrystalline p-i-n solar-cell with 8.5% conversion efficiency at AM1.5 [43]. A key attractive feature of the uc-Si:H solar-cell is that it is stable under extended light-soaking conditions. The Neuchatel group pushed the technology further by combining an a-Si top cell (1.7 eV) with a mc-Si:H bottom cell (1.0 eV) to form a tandem cell structure called "micromorph cell", and achieving 13.1% efficiency (12% stabilized efficiency). The parameters that contribute to the performance of the micromorph tandem cells are far from being optimized, and there is considerable room for further improvement of this promising technology. With its promise for higher efficiency, lower cost, and better stability, this technology could truly become the next generation of thin-film solar cells. The current status and future opportunities of solar cells based on this technology are discussed in a recent review article [44].

2.6.3 Thin-Film GaAs Solar Cells

GaAs is one of the III-V semiconductors that have an optimum direct bandgap (1.43 eV), high absorption coefficient, and high carrier mobilities, making it an ideal material for solar-cell applications. This is borne out by the recent achievement of extremely high efficiencies, in the range of 25–32%, in single- and multijunction devices made of GaAs and its alloys grown epitaxially on GaAs single-crystal substrates [45–47]. These cells and modules are now produced commercially for space applications. However, for terrestrial solar cells, it is desirable to make thin-film devices at a cost that is competitive with other advanced thin-film options. One of the options is to grow GaAs thin film on low-cost substrates such as glass, metals, or ceramics. Unfortunately, this yields materials with small grain size, high thermal stress, high dislocation densities, and poor electronic properties that adversely impact solar-cell efficiency. In spite of these limitations, a moderate level of efficiencies (9–12%) in thin-film GaAs deposited on low-cost substrates was achieved in the early 1980s. Further progress has been made in recent years in GaAs thin solar cells by using two approaches: (1) reusable substrates by using epitaxial lift-off techniques, and (2) growth on low-cost substrates. Thin-film GaAs/GaInP solar cells have been grown on single-crystal substrates, which are then separated from the substrate and transferred to a glass substrate. Similar approaches have been used to fabricate GaAs thin-film solar cells on a Pd-coated Si substrate, which led to 20% efficiency [48]. Efficiencies of 20–21% have been achieved on a GaAs cell grown on a submillimeter-grain-size, poly-Ge substrate [49]. Most recently, Motorola has announced the ability to grow high-quality GaAs thin film epitaxially on a Si substrate by using a thin $SrTiO_3$ as an intervening layer. The formation of an ultrathin (~ 8 Å) SiO_2 layer at the interface allows for subsequent strain relaxations of the GaAs layer. This important development opens up the possibility of growing multijunction III-V solar cells on relatively inexpensive Si substrates. In view of these developments, it seems probable that high-efficiency devices of GaAs thin-film solar cells can be grown in a cost-effective way provided some fundamental issues related to growth and defect passivation, grain boundaries, and dislocations are solved.

2.6.4 Dye-Sensitized TiO_2 Thin-Film Solar Cells

The idea of dye sensitization of inorganic materials had been around for a long time, and a vast amount of literature exists on this subject. However, the first use of dye-sensitized TiO_2 photoelectrochemical cells for solar-energy conversion was demonstrated by Deb et al. [50] in the mid-1970s and several U.S. patents were issued to them in 1978 on this topical area. TiO_2 fine particles and thin films, both in anatase and rutile form, were used. N-methylphenazinium dye was used as a sensitizer to extend the wavelength response in the visible region of the solar spectrum. The conversion efficiencies of such a device was low, and photostability of the dye was an issue. A breakthrough occurred in recent years when O'Regan and Graetzel [51] reported a photoelectrochemical solar cell in which a thin film based on nanoparticles of TiO_2 was sensitized by a more efficient and stable dye system based on Ru(11)-complexes.

Typically, the device structure, shown in Fig. 2.11a, consists of a 10–20-μm-thick film of nanocrystalline TiO_2 particles (10–30 nm in diameter), preferably in anatase form, that contains a monolayer of chemisorbed dye molecules that are coated on a transparent conducting glass. The TiO_2 is highly porous (50% porosity) and its pores are filled with a nonaqueous electrolyte containing a II-III redox couple. A transparent counter electrode consists of a thin layer of Pl-coated conducting glass, which is edge sealed to complete the device. The most efficient dye found to date, the so-called "black dye", consists of 4,9,14-tricarboxy 2,2'-6,6'-terpyridyl ruthenium (ii)trithiocyanate complex, which strongly adsorbs in the wavelength range of 400–900 nm. The highest efficiency in a 0.294-cm^2 area device is 10.96%. The most recent development

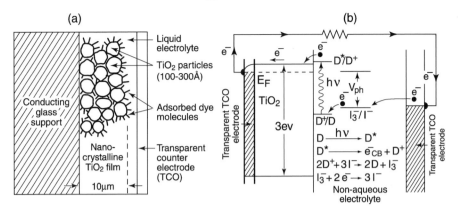

Fig. 2.11. Dye-sensitized TiO$_2$ solar-cell: (a) device structure, (b) operational principle

and status of dye-sensitized TiO$_2$ solar cells is reviewed in a recent paper [52]. From a device point of view, it would be desirable to replace the liquid electrolyte in this system with a solid-state material, preferably with a transparent p-type semiconductor. Initial results appear very promising.

The dye-sensitized TiO$_2$ solar cell offers a significant cost advantage over conventional solid-state PV devices. This is because of the relatively inexpensive component materials and inherently simple device processing. The achievement of 11% efficiency in small-area devices at the Ecole Polytechnics Federal of Lausanne (EPFL) has generated considerable interest worldwide for the commercialization of this technology. Both EPFL and the Institute of Photovoltaics (INAP) in Germany have made significant progress in the design and fabrication of large-area interconnected modules [53]. INAP has successfully fabricated a 112.3-cm^2 module of 12 interconnected cells and achieved 7.1% efficiency. Meanwhile, EPFL has reported that testing this device for more than 7000 h of continuous operation showed no degradation, which removes a lingering doubt about its long-term stability [54].

Some of the key research issues that are currently being addressed include design and synthesis of new dyes with improved response characteristics, dynamics of electron transfer and recombination processes, surface and interface properties including surface modification, charge-transport processes in nanostructure TiO$_2$ films, role of crystal structure, such as an anatase versus rutile modifications, and the replacement of the liquid electrolyte by solid-state materials.

2.6.5 Novel Ternary and Multinary Compounds

The ternary and quaternary semiconductors such as CIS and CIGS, as discussed above, have turned out to be highly efficient and stable PV materials as compared to their II-VI binary analogs. The immense success with CIS-based ternary material for PV opens almost limitless possibilities of finding other ternary and multinary compounds that could be even better than CIS. In fact, from bandgap consideration alone, CIS (energy gap = 1.04 eV) would not have been considered as the most promising material for PV. One can, with reasonable certainty, make a prediction that the next generation of PV materials is likely to come from this class of ternary and multinary compounds. Among the many different ternary compounds, only those belonging to a class of diamond-like compounds are likely to be more promising because of their close analogy with traditional semiconductors with respect to their composition, crystal structure, the nature of chemical bonds, and therefore, electronic structure. Many of these materials crystallize in the tetragonal chalcopyrite structure. The vast majority of these compounds are valence compounds with tetrahedral structure due to sp^3 hybridization of their valence electrons, such as the elemental semiconductors, silicon, and germanium.

There are also many ternary compounds that show s^2p^6 hybridization, which leads to octahedral phases. These materials in the tetrahedral and octahedral phases, with an average of four electrons per atom, form a class of valence compounds that show interesting semiconducting properties appropriate for PV application. This class of materials was extensively studies in the 1960s by a Russian group led by Goryunova [55], particularly with regard to their crystal chemistry, electronic structure, and semiconducting properties. An excellent monograph on ternary diamond-like semiconductors gives the status of the field until 1969 [56]. Many of the compounds have optical and electronic properties that may lead to excellent PV materials. The representative classes of two cation ternary compounds with four electrons per atom are shown in Table 2.5. They are classified under two categories, those with and without these defects.

Table 2.5. Representative classes of two-cation ternary compounds

Ternary compounds	
Without defect	With defect
$A^IB^{III}C_2^{VI}$ (CuInSe$_2$)	$A_2^IB^{II}C_4^{VII}$ (Ag$_2$HgI$_4$)
$A_2^IB^{IV}C_3^{VI}$ (Cu$_2$GeSe$_3$)	$A^{II}B_2^{III}C_4^{VI}$ (CdIn$_2$Te$_4$)
$A_3^IB^VC_4^{VI}$ (Cu$_3$AsSe$_4$)	$A_2^IB^{IV}C_4^{VI}$ (Zn$_2$GeSe$_4$)
$A^IB_2^{IV}C_3^V$ (CuGe$_2$P$_3$)	$A^{II}B^{IV}C_3^{VI}$ (ZnGeSe$_3$)
$A^{II}B^{IV}C_2^V$ (ZnSiP$_2$)	$A_3^{III}B^VC_3^{VI}$ (Ga$_3$PSe$_3$)

There are about 80 compounds from among the five classes of ternaries without defects that are likely to exist, a large fraction of which could be potential PV materials. Similarly, many more compounds with defects also exist and could be of potential interest for PV. If one makes quaternary alloys such as CIGS from these ternaries, the list can grow quite lengthy.

The apparent compositional complexity of ternary and quaternary materials makes one conclude that it will be difficult to synthesize and control the optoelectronic properties of these compounds. On the contrary, the ternary compounds offer several advantages over their binary analogs: (1) relatively lower melting points make them easier to prepare, (2) low-temperature processing reduces contamination during synthesis, (3) they offer a wider choice of bandgap engineering by compositional changes and ordering in cation or anion sublattices, and (4) there is a greater possibility of finding the optimum PV material with respect to cost, efficiency, stability, and environmental sensitivity. Of course, it is fairly easy to predict which materials should be attractive for PV from their optoelectronic properties, but the development of a technology based on a given material takes years of effort.

2.6.6 Organic Solar Cells

During the past two decades, major progress has been made in fabricating multilayer organic thin-film devices for a variety of applications such as light-emitting diodes (LED) and thin-film transistors (TFT) [57]. In recent years, spectacular advances have been made, particularly in developing stable and efficient LEDs that are approaching the performance of their inorganic counterpart. These developments have raised enormous interest in fabricating flexible, low-cost, and efficient thin-film PV devices based on organic materials that might eventually rival the performance of more advanced inorganic thin-film solar cells. Considerable research has been done on organic solar cells during the past two decades; basically, two device configurations have been explored. These are: (1) Schottky-type devices in which the organic light-absorbing layer is sandwiched between two metal electrodes of asymmetric (high and low) work functions, and (2) a bilayer device in which a heterojunction is formed between electron doners and acceptors. The PV performance of most of these devices has been very poor $(0 < 1.0\%)$, and it is only during the past few years that the efficiency has been increased to around 3.6%. Another interesting, recent development has been the fabrication of a thin-film PV structure from a solution consisting of a self-organized discotic liquid crystal and a crystalline conjugated material. This device has shown 2% conversion efficiency [58].

The recent advances in solar-cell efficiency of a few representative organic solar cells are given in Table 2.6. Clearly, the major improvement in efficiency has been achieved by fabricating devices with a composite of electron-donor-type conjugated polymers with electron-acceptor-like fullerene derivatives. Major improvement in the efficiencies reported in this table are mostly

Table 2.6. Reported efficiencies of some recent organic solar cells at AM1.5

Device structure	η (%)	Ref.
ITO/CuP$_c$/PV/Ag	1.0	59
ITO/MEH-PPV-C$_{60}$/Al or Ca	2.9	60
ITO/POPT + MEH - CN - PPV (5%)/top layer MEH-CN-PPV + POPT) 5%/Al or Ca/bottom layer	1.9	61
Ag/ITO/pentacene/Al or Mg	2.4	62
ITO/CuP$_c$/PTCBI/BDP:PTCBI/Ag	2.4	63
ITO/PEDOT/MDMO-PPV: PCBM/LiF/Al	2.5	64
ITO/PEDOT: PSS/CuP$_c$/C$_{60}$/BCP/Al	3.6	65
ITO/PEDOT:PSS/MDMO-PPV: PCBM/Al	4.3	66

CuP$_c$: copper phthalocyanine; PEDOT: polyethylenedioxythiophene;
PSS: polystyrenesulfonate; C$_{60}$: fullerene; BCP: bathocuproine;
PTCBI: perylenetetracarboxylic-bis-benzimidazole;
MEH: methoxyethoxyhexyloxy; PPV: polyphenylenevinylene;
POPT: derivative of polythiophene

on small-area devices (0.1 cm^2). However, the group at the University of Linz, Netherlands, has fabricated conjugated polymer-based large-area devices (10 × 10 cm) on flexible ITO-coated plastic substrates and has routinely achieved 2% efficiency. A recent review article [67] gives an excellent account of the recent developments in conjugated polymer-based solar cells including the underlying photophysics.

The poor performance of organic solar cells is primarily attributed to the following two main factors: (1) inefficient photoinduced charge generation due to low exciton diffusion length as compared to the optical absorption length, and (2) poor collection efficiency due to very low carrier mobilities ($\sim 10^{-3}$ cm^2/Vs). However, carrier mobilities approaching those of amorphous silicon have been achieved in certain organic semiconductors used for TFT applications. Continued progress in this area will depend on materials improvement and innovative device engineering to optimize these two parameters. Judging from the rapid progress made in recent years, the future outlook for low-cost organic solar cells looks bright.

2.6.7 Novel Approaches to High-Efficiency Thin-Film Solar Cells

The multijunction approach has significantly increased the efficiency over single-junction devices in a-Si:H- and GaAs-based solar cells. In principle, it is possible to increase the efficiency close to thermodynamic limits by stacking a large number of cells in tandem. From the economic perspective, this approach quickly runs into the law of diminishing returns. During the past two decades, several innovative approaches have been proposed that get around this problem, and these are: hot-carrier solar cells, quantum wells, and quantum-dot devices. These ideas are based on the principle of creating multiple electron–hole pairs per absorbed photon by utilizing the excess kinetic energy of electrons and holes, provided the extremely fast and completing process of carrier relaxation by phonon emission can be taken care of. It has been shown that the carrier relaxation time can be greatly increased from the femtosecond to the nanosecond regime by using quantum-well superlattice structure and quantum dots (QDs). In QDs, the photocurrent can be enhanced significantly by two primary processes: carrier multiplication by the inverse Auger process, and higher carrier mobilities due to miniband formation. The quantization also offers other advantages like: bandgap modulation, enhanced optical absorption, and indirect-to-direct bandgap formation, which can be used for more efficient photoconversion processes. Another interesting idea for carrier multiplication involves intermediate band formation [68]. These ideas have the potential to increase the efficiency over current solar cells by 2–3 times. Whether any of these approaches can lead to a quantum jump in thin-film solar-cell efficiency in the future remains highly speculative. Several recent papers have discussed these ideas at length [69–71].

Acknowledgments

This work was supported under US/DOE contract #DE-AC36-98-GO10337. My thanks go to Melody Mountz, Inger Jafari, Susan Moon, and Al Hicks for assistance in preparing the manuscript.

References

1. Yang, J., Banerjee, A., Guha, S. (1997): *Appl. Phys. Lett.* 70:2975.
2. Yang, J., Banerjee, A., Glatfelter, T., Sugiyama, S., Guha, S. (1997): Proc. 26th IEEE PVSC., p.563.
3. Yang, J., Banerjee, A., Lord, K., Guha, S. (2000): Proc. IEEE PVSC., p.742.
4. Nishiwaki, H., Sakai, S., Haku, H., Ohnishi, M. (1995): *Prog. Photovolt.* 3:221.
5. Shah, A., Dutta, J., Wyrsch, N., Prasad, K., Curtius, H., Finger, F., Howling, A., Hollenstein, Ch. (1992): *Proc. Mater. Res. Soc. Symp.* 258:15.
6. Chatam, H., Bhat, P.K. (1989): *Proc. Mater. Res. Soc. Symp.* 149:447.
7. Jones, S.J., Liu T., Deng X., Izu M. (2000): Proc IEEE PVSC, p.845.
8. Nelson, B.P., Xu, Y., Mahan, A.H., Williamson, D.L., Crandall, R.S. (2000): *Proc. Mater. Res. Soc. Symp.* 609:A 22.8.1.
9. Wang, Q., Iwaniczko, E., Xu, Y., Nelson, B.P., Mahan, A.H., Crandall, R.S., Branz H.M. (2000): Proc. IEEE PVSC.
10. Mori, Y., Yoshii, K., Yasutake, K., Kakiuchi, H., Domoto, Y., Tarui, H., Kiyama, S. (1999): 9th Int. Conf. on Precision Engineering, 537, The Japan Society of Precision Engineering, Tokyo.
11. Okamoto, S., Terakawa, A., Maruyama, E., Shinohara, W., Hashikawa, Y., Kiyama, S. (2000): Proc. IEEE PVSC, p.736.

12. Futako, W., Fukutani, K., Kannabe, M., Kamiya, T., Fortmann, C.M., Shimizu I. (1997): Proc. 26th IEEE PVSC, p.581.
13. Futako, W., Shimizu, J., Fortmann, C.M. (1996): *Proc. Mater. Res. Soc.* 420:431.
14. von Roedern, B., del Cueto, J.A. (2000): *Proc. Mater. Res. Soc. Symp.* 609:A10.4.
15. Schropp, R.E.I., Zeeman, M. (1983): Amorphous and Microcrystalline Solar Cells, Kluwer Academic Publishers, Dordrecht.
16. Adler, D. (1983): *Solar Cells* 9:133.
17. Stutzman, M., Jackson, W.B., Tsai, C.C. (1985): *Phys. Rev.* B 32:23.
18. Branz, H.M. (1999): Phys. Rev. B 59:5498.
19. Biswas, R., Pau, B.C. (1998): *Appl. Phys. Lett.* 72:371.
20. Cheong, H.M., Lee, S.H., Nelson, B.P., Mascarenhas, A., Deb S.K. (2000): *Appl. Phys. Lett.* 77:2686.
21. Arya, R.R., Bennett, M., Lin, B., Willing, F., Newton, J., Ganguly, G., Liu, S. (2000): Proc IEEE PVSC, p.1433.
22. Gulia, S., Yang, J., Banerjee, A., Hoffman, K., Sugiyama, S. (USSC), Jones, S., Deng, X., Doehler, J., Szu, M., Ovskinsky, H.C. (ECD) (1997): Proc. 26th IEEE PVSC, p.607.
23. Wagner, S., Carlson, D.E., Branz, H.M. (1999): *Proc. Electrochem. Soc.* 99-11:219.
24. Cusano, D. (1963): Solid State Electron 6:217.
25. Farhenbruch, A.L., Bube, R.L. (1983) Fundamentals of Solar Cells, Academic Press, New York, p.489.
26. Okeyama, H., Aramato, T., Kumazawa, S., Higuchi, H., Arita, T., Shibutani, S., Nishio, T., Nakajima, J., Tsuji, M., Hanafusa, A., Hibino, T., Omura, K. (1997): Proc. 26th IEEE PVSC, p.343.
27. Britt, J., Ferekides C. (1993): *Appl. Phys. Lett.* 62:2851.
28. Wu, X., Ribelin, R., Dhere, R.G., Albin, D.S., Gessert, T.A., Asher, S., Levi, D.H., Mason, A., Moutino, H.R., Sheldon, P. (2000): Proc. 28th IEEE PVSC, p.470.
29. Rose, D., Powell, R., Jayamaha, U., Maltby, M., Giolando, D., McMaster, A., Kormanyos, K., Faykosh, G., Klopping, J., Dorer, G. (2000): Proc. IEEE PVSC, p.428.
30. Bonnet, D., Harr, M. (1998) Proc. 2nd World Conf. PVSEC.
31. Compann, A.D., Sites, J.R., Birkmire, R.W., Ferekides, C.S., Fahrenbruch, A.L (1999): *Proc. Electrochem. Soc.* 99-11:241.
32. Shay, J.L., Wagner, S., Kasper, H.M. (1975): *Appl. Phys. Lett.* 3:3.
33. Kazmerski, L.L., White, F.R., Morgan, A.K. (1976): *Appl. Phys. Lett.* 29:268.
34. Devaney, W.E., Mickelsen, R.A., Chen, W.S. (1985): Proc. 18th IEEE PVSC. p.1733.
35. Contreras, M., Egass, B., Ramanathan, K., Hiltner, J., Swartzlander, A., Hasoon, F., Noufi, R. (1999): *Prog. Photovolt: Res. Appl.* 7:311.
36. Contreras, M.A., Tuttle, R., Gabor, A., Tennant, A., Ramanathan, K., Asher, S., Franz, A., Keane, J., Wang L., Scofield ,J., Noufi, R. (1994): Proc. 24th IEEE PVSC., p.68.
37. Rockett, A., Bhattacharya, R.N., Eberspacher, C., Kapur, V., Wei, S.H. (1999): *Proc. Electrochem. Soc.* 99-11:232.

38. Photovoltaics for the 21st Century., ed by Kapur V.K., Mcconnell R.D., Carlson, D., Ceasar, G.P., Rohatgi, A. (1999): *Proc. Electrochem. Soc.* 99-11.
39. Photovoltaics for the 21st Century II, ed by McConnell, R.D., Kapur, V.K.: (2001): *Proc. Electrochem. Soc.* 2001-10.
40. Future Generation Photovoltaic Technologies, ed. by McConnell, R.D. (1997): Proc. AIP Conf. 404.
41. Imazuni, M., Ito, T., Yamaguchi, M., Kaneko, K. (1997): *J. Appl. Phys.* 81:7635.
42. Deb, S.K., Sopori, B. (2000): Handbook of Thin Film Devices, ed. Francombe M.H., Semiconductor Optical and Electro-Optical Devices, Chap. 11, 2, 311-362, Academic Press.
43. Meier, J., Keppner, H., DuBaie, S., Kroll, U., Platz, R., Torres, P., Pennet P., Ziegler, UY., Selvan, J.A.A., Cuperus, J., Fischer, D., Shah, A. (1998): *Proc. Mater. Res. Soc. Symp.* 507:139.
44. Keppner, H., Meier, J., Torres, P., Fischer, D., Shah, A. (1999): *Appl. Phys.* A 69:169.
45. Kurtz, S., Olson, J.M., Kibbler, A. (1990): Proc. IEEE PVSC, p.138.
46. Bertness, K., Kurtz, S., Friedman, D., Kibbler, A., Kramer, C., Olson, J.M. (1994): *Appl. Phys. Lett.* 65:989.
47. Takamoto, T., Ikeda, E., Kurita, H., Ohmori, M. (1997): *Appl. Phys. Lett.* 70:381.
48. Omnes, F.O., Guillaume, J.C., Jager-Walden, G., Natafa, G., Vennegues, P., Gilbert, P.(1996): IEEE Trans. Electron Devices 43:1806.
49. Venkatsubramanian, R., O'Quinn, B.C., Sivola, E., Keyes, B., Ahrenkiel, R. (1997): Proc. 26th IEEE PVSC, p.811.
50. Deb, S.K., et al. (1978): US Patents 4, 117, 210; 4, 080 ,488; 4, 084, 043; 4, 118, 546, and 4, 118, 547.
51. O'Regan, B., Graetzel, M. (1991): *Nature* 353:737.
52. McEvoy, A.J., Nazeeruddin, M.K., Rothenberger, G., Graetzel, M. (2001): *Proc. Electrochem. Soc.* 2001-10:69.
53. Chmiel, A., Gehring, J., Uhlendorf, J., Jestel, D. (1998): Proc. 2nd World PVSC, p.53.
54. Gratzel, M. (1997): AIP Conf. Proc. 404:119.
55. Goryunova, N.A. (1965): The Chemistry of Diamond-Like Semicoductors, Chapman and Hall.
56. Berger, L.L., Prochukan, V.D. (1969): Ternary Diamond-like Semiconductors, Consultant Bureau, New York London
57. Forrest, S.R. (1997): *Chem. Rev.* 97:1793.
58. Schmidt-Mendi, L., Fechtenkotter, A., Mullen, K., Moons, E., Friend, R.J., Mackenjee, J.D. (2001): *Science* 293:1119.
59. Tang, C.W. (1986): *Appl. Phys. Lett.* 48:183.
60. Yu, G., Gao, J., Hummelen, J.C., Wudl, F., Heager, A.J. (1995): *Science* 270:1789.
61. Granstrom, M., Petritsch, K., Arias, A.C., Lux, A., Andersson, M.R., Friend, R.H. (1998): *Nature* 395:257.
62. Schon, J.H., Kloc, Ch., Bucher, E., Batlogg, B. (2000): *Nature* 403:408.
63. Peumans, P., Bulovic, V., Forest, S.R. (2000): *Appl. Phys. Lett.* 76:2650.

64. Shaheen, S.E., Brabec, C.J., Sariciftci, N.S., Padinger, F., Fromherz, T., Hummelen, J.C. (2001): *Appl. Phys. Lett.* 78:841.
65. Peumans, P., Forest, S.R. (2001): *Appl. Phys. Lett.* 79:126.
66. Shaheen, S.E., Radspinner, R., Peyghambarian, N., Jabbour G.E. (2001): *Appl. Phys. Lett.* 79:2996.
67. Brabec, C.J., Sariciftci, N.S., Hummelen, J.C. (2001): *Adv. Funct. Mater.* 11:15.
68. Marti, A., Cuadra, L., Luque, A. (2001): *Proc. Electrochem. Soc.* 2001-10:46.
69. Green, M.A. (2001): *Proc. Electrochem. Soc.* 2001-10:3.
70. Nozik, A.J. (2001): *Proc. Electrochem. Soc.* 2001-10:61.
71. Barnham, K.W.J., et al. (2001): *Proc. Electrochem. Soc.* 2001-10:30.

3

Electrical and Optical Properties of Amorphous Silicon and Its Alloys

Hiroaki Okamoto

This chapter will concern itself with a detailed description of the basic concepts of electrical and optical properties of amorphous semiconductors with respect to the specific example of amorphous silicon. Using a simplified physical model, we will discuss how the electronic states in the vicinity of the band edge, and the optical and electronic processes associated with these, are affected by the structural disorder. An attempt will be made to interpret and explain the various physical properties of hydrogenated amorphous silicon (a-Si:H) including the photoinduced changes.

3.1 Simplistic Model for Band-Edge Electronic Properties

3.1.1 Fundamental Aspects Near the Mobility Edge

3.1.1.1 Density-of-States in the Band-Edge Region

In this section, the band-edge structure in a system involving weak disorder is examined within the context of a site-diagonal disordered simple tight-binding model. The electronic system is supposed to be described by a linear combination of Wannier states $|i\rangle$ localized at regularly arranged atomic or molecular orbital sites [1, 2]. If the site energy is assumed to take a random value v_i, and the transfer energy between the nearest-neighbor sites is constant at V, then the Hamiltonian is given by

$$H = V \sum_{i \neq j} |i\rangle\langle j| + \sum_i |i\rangle\langle i| \ . \tag{3.1}$$

As is well known, the off-diagonal term gives the energy dispersion relationship of the regular lattice system $\epsilon = \epsilon_0(k)$, and the width of the created band B is represented by $2z|V|$, where z is the coordination number. Let us assume that the diagonal component v_i of the second term follows a Gaussian distribution with zero mean and variance W^2, and consider this as the perturbation term of disorder for the regular system presented by the first term in (3.1).

A rough sketch of the density-of-states (DOS) spectrum, $D(\epsilon)$, in the vicinity of the band edge is given according to the renormalized second-order perturbation method. It is now assumed that the DOS of a nonperturbed system is proportional to $\epsilon^{1/2}$ near the band edge [3].

$$D(\epsilon) = \frac{1}{2\sqrt{2}\,\pi^2}(Va^2)^{-3/2}\left(\sqrt{(\epsilon - \sigma_r)2 + \sigma_i^2} + \epsilon - \sigma_r\right)^{1/2}, \qquad (3.2)$$

where a denotes the interatomic distance, σ_r a negative value close to $-4W^2/B$ which corresponds to the shift in the DOS spectrum, and σ_i is the imaginary part of the self-energy representing "blur" of the spectrum, being proportional to $W^2V^{-3/2}E^{1/2}$. From (3.2), it is evident that for the energy region $\epsilon > \sigma_r + 4W^3/B^2$, the DOS resembles that of a nonperturbed system shifted by σ_r. While it is difficult to show explicitly by using such a simple analysis, it is expected that a band tail of the exponential form [2]

$$D(\epsilon) \propto \exp\left(\frac{\epsilon}{E_{bu}}\right) \qquad (3.3)$$

follows the energy behavior mentioned above toward the lower-energy side, with the characteristic energy E_{bu} being proportional to W^2/B.

The basic concept of the above discussion is that the electron state of a disordered system is composed of superimposed Bloch states of its regular counterpart. The mean free path Λ constitutes an index for the proportion of included **k**-states, and is give approximately by

$$\Lambda \cong 4\pi\sigma\left(\frac{V}{W}\right)^2 \qquad (3.4)$$

for $E > \sigma_r$ [2]. The range of k to be mixed is of the order of $1/\Lambda$ and its center of gravity $\langle k \rangle$ is related to $D(\epsilon)$ by $\langle k \rangle \propto Va^2D(\epsilon)$. In consideration of the effective mass m^* defined at the band edge of a non-perturbed system, which is given in the simple tight-binding model by $\hbar^2/2Va^2$, it seems that even in a disordered system the same effective mass would be assigned for $\epsilon > \sigma_r + 4W^3/B^2$ [3].

3.1.1.2 Conductivity Spectrum and Mobility Edge

Let us start here from what is known as the Kawabata expression for the energy (ϵ) and size (L)-dependent conductivity, which is derived from a microscopic linear response theory due to Kubo [2, 3]:

$$\sigma_L(\epsilon) = \sigma_B(\epsilon) + \Delta\sigma_L(\epsilon) . \qquad (3.5)$$

The first term, $\sigma_B(\epsilon)$, represents the classical Boltzmann conductivity, and the second term, $\Delta\sigma_L(\epsilon)$, a quantum-correction term reflecting the multiple scattering effect due to the random scattering potential, that is, the site-diagonal component in (3.1). The first term is written simply as

$$\sigma_B(\epsilon) = \frac{1}{\pi^3} \frac{e^2}{\hbar \Lambda} u \ ,$$ (3.6)

where u is the reduced energy $u = (\epsilon - \sigma_r)/\Delta_c$ with Δ_c defined by

$$\Delta_c = \frac{1}{16\pi^3} \frac{W^4}{V^3} \ .$$ (3.7)

In accordance with the standard Green's function diagram technique, the second term of (3.5) is evaluated to be

$$\Delta\sigma_L(\epsilon) = -\sigma_B(\epsilon) \cdot \frac{1}{u}\left(1 - \frac{\Lambda}{L}\right) \ .$$ (3.8)

if the first-order quantum correction term involving $1/u$ is taken into account. Accordingly, the conductivity spectrum is given as the sum of (3.6) and (3.8):

$$\sigma_L(\epsilon) = \frac{1}{\pi^3} \frac{e^2}{\hbar}\left(\frac{1}{\xi} + \frac{1}{L}\right) \ ,$$ (3.9)

where ξ represents a correlation length defined by $\xi = \Lambda/(u-1)$. Equation (3.9) is believed to be correct so far as no factors either breaking the time-reversal symmetry of the electron system (magnetic field or spin-dependent scattering) or distorting the coherence of the electron wave (inelastic scattering) are involved.

Now, let us examine (3.9) in some detail. If the sample size L is comparable to the mean free path λ, the correction term $\Delta\sigma_L(\epsilon)$ disappears, leaving only the classical Boltzmann term. It can be clearly seen that as the sample size grows, then the conductivity decreases in inverse proportion to L, and when it exceeds the correlation length ξ approaches a size-independent value $(u-1)/\Lambda$, which governs the electrical conductivity in an actual infinitely large sample, that is,

$$\sigma(\epsilon) = \frac{1}{\pi^3} \frac{e^2}{\hbar \Lambda}(u-1); \qquad u \geq 1$$
$$= 0; \qquad\qquad\qquad u < 1$$ (3.10)

for $L \rightarrow \infty$. Equation (3.10) indicates that the electrical current flows only in energy states that meet $u > 1$ or, in other words, $\epsilon > \Delta_c$. The energy position at this boundary defines the mobility edge ϵ_c:

$$\epsilon_c = -4\frac{W^2}{B} + \frac{1}{16\pi^3} \frac{W^4}{V^3} \ ,$$ (3.11)

where the energy is referred to the band-edge position of the nonperturbed regular system. The energy states higher than this critical value have extended wave functions, and states with lower energies are localized states that are characterized by a finite localization length being around -2ξ. Even in the extended states, the amplitude of the wave function appears to be fluctuating for the energy range in which $\xi > L > \Lambda$ holds; i.e., $1 < u < 2$, since, as mentioned above, the size-independent conductivity is established for $L \gg \xi$.

It should be noted here that the electrical conductivity observed for an actual sample (with the sample size L infinitely large in comparison with the mean free path Λ) is expressed by using the Fermi–Dirac distribution function $f(\epsilon)$ as

$$\sigma(T) = \int \sigma(\epsilon) \left(-\frac{\partial f(\epsilon)}{\partial \epsilon} \right) d\epsilon \ . \tag{3.12}$$

Putting (3.10) into (3.11) leads to a simple expression of the electrical conductivity for disordered nondegenerate semiconductors:

$$\sigma(T) = \frac{1}{3\pi^2} \frac{e^2}{\hbar a} \left(\frac{\Lambda k_{\mathrm{B}} T}{aV} \right) \exp \left(\frac{\epsilon_F - \epsilon_c}{k_{\mathrm{B}} T} \right) \ , \tag{3.13}$$

where ϵ_F denotes the Fermi energy.

3.1.2 Optical Absorption Spectrum

3.1.2.1 Phenomenological Scaling of Optical Absorption

A remarkable similarity holds in the optical absorption edge spectrum for various amorphous semiconductor systems: an exponential rise (Urbach region) followed by a slowly varying regime (Tauc region). The observation is probably a manifestation of the presence of a universal absorption line shape, $L(E)$, inherent to moderately disordered materials. To explore this problem, we first introduce the dimensionless photon energy χ by $(E - E_o)/\Gamma$, where E_o denotes the optical bandgap, and Γ is a physical quantity reflecting the microscopic structure of the material, although not specified at this point. It is then straightforward to show that the normalized absorption, g, defined by $L(E)/L(E_o)$ obeys [4]

$$\chi \frac{\mathrm{d} \ln g(\chi)}{\mathrm{d}\chi} = n_T \qquad \text{for } g \gg 1 \ ,$$
$$\qquad\qquad = \ln g(\chi) \text{ for } g \ll 1 \ , \tag{3.14}$$

where $g \gg 1$ refers to the Tauc regime (the Tauc exponent, $n_T \simeq 2 - 3$) and $g \ll 1$ to the Urbach regime, in accordance with the general experimental absorption features of amorphous semiconductors [5]. It is important here that the right-hand side of (3.14) does not include χ explicitly, but is defined by g alone, which we write as $\beta(g)$. A possible form of the $\beta(g)$ function is represented by

$$\beta(g) = \ln g - \ln \left(1 + a \frac{(g-1)^2}{g+b} \right) \ , \tag{3.15}$$

in which a and b are postulated to be independent of any material parameters.

The scaling/renormalization group procedure presented above does not provide practical values for the constants a and b in (3.15). They must be determined through microscopic arguments based on a microscopic model as given in (3.1). In terms of the perturbative time-dependent Green's function formalism for the frequency-dependent conductivity, a and b appear to be expressed as a function of

$$\Delta = (E_x - E_o)/\Gamma \tag{3.16}$$

and

$$\delta = \frac{1}{2} \frac{(W_c^2 + W_v^2)^{3/2}}{W_c^2 B^c + W_v^2 B^v} \ , \tag{3.17}$$

where E_x denotes the bandgap of the nonperturbed regular system. The energy scaling parameter Γ is now identified as

$$\Gamma = \frac{1}{2} \frac{(W_c^2 + W_v^2)^2}{W_c^2 B^c + W_v^2 B^v} \ . \tag{3.18}$$

The parameter Δ is determined for each δ so as to maintain a and b constant, yielding a relation:

$$E_o = E_x - 6.25 E_u \ , \tag{3.19}$$

where E_u denotes the Urbach energy represented by

$$E_u = 1.3\delta^{-1.2} \frac{(W_c^2 + W_v^2)^2}{W_c^2 B^c + W_v^2 B^v} \ , \tag{3.20}$$

which is valid for δ in the range from 0.1 to 0.3. At the same time, constants a and b turn out to be 0.0628 and 0.4183, respectively, and the Tauc exponent $n_T = -\ln a$ turns out to be about 2.77, being very close to 3, rather than the widely used value 2. There are several pieces of experimental evidence supporting $n_T \simeq 3$, whereas no theoretical reasoning has been provided so far. The theoretical prediction (3.19) agrees quantitatively with Cody's empirical law $E_o = E_x - 6.2 E_u$ [5] for a-Si:H. On the other hand, it is usually stated that E_u reflects the broader tail of either of the bands involved in optical transitions; W_μ^2/B^μ ($\mu = c$ or v). However, this condition may not always be true as is easily recognized from (3.20) even if the contribution from δ is ignored.

It would be interesting to see whether the predicted absorption line shape, an integrated form of (3.15), represents the actual absorption spectrum. Examples of such comparison are displayed in Fig. 3.1, in which the energy scaling parameter Γ is replaced with E_u according to (3.20). Quite nice fits are achieved by the scaling of the horizontal axis using experimentally determined E_o ($n_T = 2.77$) and E_u, whereas the fits have failed when using E_o ($n_T = 2$), both inferring the plausibility of the phenomenological scaling approach demonstrated in this section.

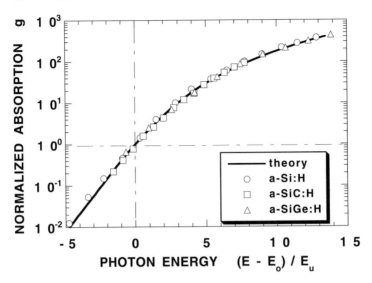

Fig. 3.1. Normalized absorption g plotted against reduced photon energy, and experimental data on a-Si:H alloys

3.1.2.2 Relation Between the Mobility Gap and Optical Gap

For amorphous semiconductors, it is quite difficult to evaluate the mobility gap which plays a central role in determining electrical properties, instead the optical band gap E_o (E_{o2} for $n_T = 2$, and E_{o3} for $n_T = 3$) is employed to specify the materials. It is therefor instructive to examine the relation between the mobility gap and optical gap although any quantitative description is not possible within the context of the simplistic TB scheme. The numerical coefficients in (3.11) and (3.20) may depend quite sensitively on the details of the bands and in particular on the size of the gap relative to the bandwidths. In order to accommodate our simplistic TB model to the realistic a-Si:H alloys, we let $2z = 75$ in accordance with the results of detailed theoretical analysis [2].

Figure 3.2 illustrates how various bandgap-related quantities change with the conduction-band disorder energy W_c. E_m denotes the mobility gap, E_{02} and E_{03} the optical bandgaps, E_g the DOS bandgap and E_u the optical Urbach energy. In the calculation, the valence-band disorder energy W_v is set at $(5/3)^{1/2} W_c (V_v/V_c)$ of which the physical meaning will be made clear in later sections, and other parameters, like V_c, V_v, and E_x, are chosen as indicated in the figure. The range of W_c in the shaded area is likely to correspond to the realistic "device quality" hydrogenated amorphous silicon. Much can be learned from this figure, however, it is left to the readers because of the pressure for an allotted space. At the end of this section, we have to note that the reference energy E_x is essentially dependent both on the material composition (hydrogen and other alloy elements) and temperature T, and the disorder

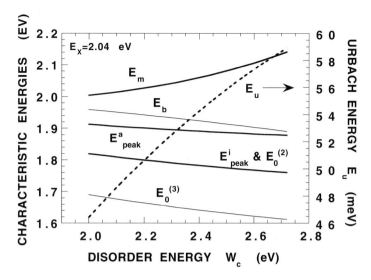

Fig. 3.2. Band-edge characteristic energies as a function of the structural disorder

energy W_μ would be temperature dependent. The latter might be visualized as

$$W^\mu(T)^2 = W^\mu(0)^2 + c^\mu \theta \coth\left(\frac{\theta}{2T}\right) , \qquad (3.21)$$

where $W_\mu(0)$ refers to the zero-temperature static disorder, c_μ denotes a quantity related to the magnitude of the deformation potential, and q the temperature converted from the effective Einstein frequency.

3.1.2.3 Physical Implication from Polarized Electroabsorption

An electric field applied to semiconductors alters the optical absorption spectrum in the vicinity of the fundamental absorption edge. For crystalline materials, the electroabsorption (EA) effect is well identified as arising from a field-induced mixing of the unperturbed one-electron Bloch functions occurring selectively near the originally vertical allowed transitions. The field-induced mixing is equivalent to spreading sharp vertical transitions over a finite range of initial and final momenta. Incorporating static disorder yields analogous but more profound effects, and provided it is sufficiently extensive, gives rise to an apparent breakdown of the momentum selection rule, as is often relied on to explain optical features of amorphous semiconductors. Since the disorder- and field-induced effects both operate in a competitive manner, any simple description of an EA spectrum is not possible for amorphous semiconductors. In this sense, EA spectroscopy has long been believed to be less informative for such materials in comparison to its successful utilization for studying the band structure of crystalline counterparts.

The situation has been changed by a recent finding of polarization-dependent EA spectra in some classes of isotropic amorphous semiconductors including amorphous silicon. The EA signal exhibits much larger strength when the field $\mathbf{F}$ is applied parallel to the polarization vector $\mathbf{e}$ of the light than when $\mathbf{F}$ is perpendicular to $\mathbf{e}$. Such an anisotropy would never be expected for macroscopically isotropic crystalline media. To understand the phenomenon as well as to explore its implication for band-edge electron states, we have developed a low-field EA theory in which the effects of disorder are incorporated in the context of a site-disordered TB model [6]. A conclusion is reached that the polarization-dependent EA effect arises essentially from a field-induced change in the matrix element for optical transitions between localized states near the band-edge, so that it serves as an indicator for the degree of disorder in amorphous semiconductors; or quantitatively aspect for the carrier mean free path.

The theory reveals that the EA signal $\Delta\alpha$, is comprised of two terms;

$$\Delta\alpha(E,\theta) \propto \left(\Delta\alpha^i(E) + \Delta\alpha^a(E)\cos^2\theta\right)F^2 , \qquad (3.22)$$

where θ denotes the angle between $\mathbf{F}$ and $\mathbf{e}$. The isotropic, $\Delta\alpha^i$, and anisotropic, $\Delta\alpha^a$, EA terms are related to the zero-field absorption $\alpha(E)$ by

$$\Delta\alpha^i(E) = \frac{\partial^3}{\partial E^3}\alpha(E) ,$$

$$\qquad (3.23)$$

$$\Delta\alpha^a(E) = \frac{49}{24}(W^cW^v)^2 \cdot \frac{\partial^7}{\partial E^7}\alpha(E) .$$

Equation (3.23) indicates that the polarization-dependent EA (PEA) occurs only when the materials involve structural disorder. As regards the EA lineshapes, a combination of (3.15) and (3.23) yields Fig. 3.3, in which the absorption α and its field-induced changes $\Delta\alpha$ are plotted against the reduced photon energy introduced in the previous section, i.e., experimental results for a device quality a-Si:H and the theoretical fits. It is found that the peak of the isotropic EA signal, $\Delta\alpha^i$, occurs at the photon energy very close to the Tauc gap ($n_T = 2$), while the anisotropic EA signal, $\Delta\alpha^a$, has its peak at a slightly higher energy. The cutoff energy of $\Delta\alpha^a$ at the higher-energy side seems to give a rough estimation of the mobility gap, although no universal identification is possible.

More important here is the ratio of the peak anisotropic and isotropic EA intensities, R_{EA}, which is related to the electron and hole mean free paths by

$$R_{EA} \equiv \frac{\Delta\alpha^a}{\Delta\alpha^i} = \frac{98}{3}\pi^2\left(\sqrt{\Lambda_c/a} + \Lambda_v/a\right)^{-4} . \qquad (3.24)$$

For device-quality a-Si:H, the R_{EA} is seen to be around 1.2 as read from Fig. 3.3. According to the same theoretical procedure, the average dipole matrix element squared R^2 for the band-edge optical transition is represented as

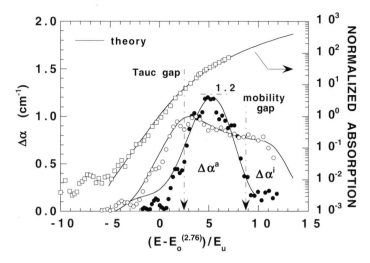

Fig. 3.3. Absorption and electroabsorption spectra of device-quality undoped a-Si:H in the annealed state

$$R^2 = 8\pi^{3/2} a^2 \left(\Lambda_c \Lambda_v / a^2 \right)^{5/4} \left(\sqrt{\Lambda_c/a} + \sqrt{\Lambda_v/a} \right)^{-5}$$
$$\times \left[1 + 2\pi \left(\sqrt{\Lambda_c/a} + \sqrt{\Lambda_v/a} \right)^{-2} \right] \tag{3.25}$$

which is experimentally found to be 10 Å^2 for various a-Si:H alloys. Combining (3.24) and (3.25) then permits us to separately evaluate electron and hole mean free paths from the measured EA polarization ratio. The story will be largely changed if either the electron or the hole mean free path is fixed by a certain experimental means, which shall be discussed in a later section dealing with the carrier-transport processes.

As for the indicator of the structural disorder, the optical Urbach energy E_u as well as the Raman TO linewidth Γ_{TO} are increased. In the final part of this section, we will briefly discuss the superiority of the EA polarization ratio, R_{EA}, as a measure of the structural disorder against the two physical quantities mentioned above. For the sake of the simplicity, we let $W_c = W_v$, by which, however, any physical significance may not be lost. The statistical variance of the bond angle is believed to be proportional to the disorder energy W, so that we have $\Gamma - 15$ cm$^{-1} \propto W^{0.5}$. On the other hand, (3.20) suggests $E_u \propto W^{0.72}$, whereas $R_{EA} \propto W^4$ as understood from (3.24). In order to examine these theoretical predictions, it would be best to demonstrate the experimental results for the correlation between E_u and R_{EA} achieved on a-Si:H formed at different growth temperatures. As found in Fig. 3.4, an 8% change in E_u is reflected in a 67% change in R_{EA}, clearly indicating that R_{EA} is much more sensitive to the structural disorder than E_u. Also plotted in this figure is the electron mobility μ_c deduced from the high-frequency modulated

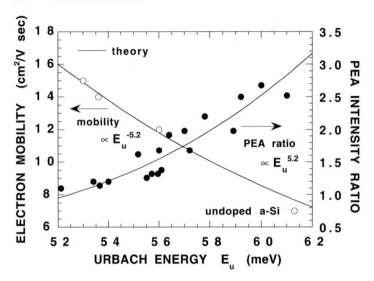

Fig. 3.4. Correlation between the electron mobility, PEA intensity ratio, and Urbach energy in undoped a-Si:H

photocurrent (MPC) spectroscopy [3]. One may see that a relation $R_{EA} \propto \mu^{-1}$ holds, of which the physical significance will be discussed later. Note here that the experimental correlations are well reproduced by the theoretical modeling in terms of physical parameters adopted to illustrate Fig. 3.2.

3.1.3 Electronic Conduction

3.1.3.1 Carrier Mobility and Mean Free Path

The macroscopic free carrier mobility μ is determined by [1]

$$\mu = \frac{\sigma(T)}{e \int \mathrm{d}\epsilon f(\epsilon) D(\epsilon)} \; , \tag{3.26}$$

in which $\sigma(T)$ is the electrical conductivity given by (3.13). Equation (3.26) is then reduced to a well-known classical form:

$$\mu = \frac{ea^2}{6k_{\mathrm{B}}T}\nu(T) \; , \tag{3.27}$$

with ν being an effective hopping frequency [3]:

$$\nu(T) = \sqrt{\frac{2}{27\pi}} \frac{V}{\hbar} \frac{\int_1^\infty \mathrm{e}^{-xu}(u-1)\mathrm{d}u}{\int_1^\infty \mathrm{e}^{-xu}\left(\sqrt{u^2 + (\pi/3)u} + u\right)^{1/2}\mathrm{d}u} \; , \tag{3.28}$$

where x is defined by $\Delta_{\mathrm{c}}/k_{\mathrm{B}}T$. It is be rather straightforward to find that the mobility is expressed in terms of the carrier mean free path Λ and $V/k_{\mathrm{B}}T$ alone if we let $ea^2/\hbar = 0.864$ cm^2 V^{-1} s^{-1} with $a = 2.35$ nm (a-Si:H).

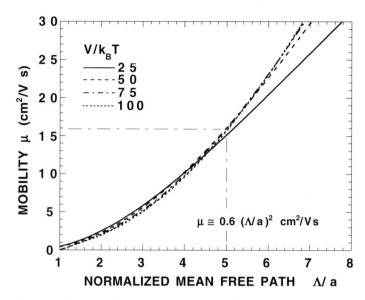

Fig. 3.5. Relation between carrier mean free path and band-edge free carrier mobility

A numerical evaluation of (3.27) leads to the relation between the carrier mobility μ and the mean free path Λ normalized by the mean atomic spacing a. The result is demonstrated in Fig. 3.5, where the mobility μ is correlated with the normalized mean free path Λ/a for various different V/k_BT, being thought to be 50–75 and 25–50 for the conduction and valence bands, respectively, in standard a-Si:H. As found in this figure, the mobility μ depends on V/k_BT very weakly for the practical range of Λ/a ($1 \leq \Lambda/a \leq 5$), so that a simple relation holds:

$$\mu \cong 0.6(\Lambda/a)^2 \quad \mathrm{cm}^2 \ \mathrm{V}^{-1} \ \mathrm{s}^{-1} \ . \tag{3.29}$$

When we recall 3.24, an approximate relationship

$$R_{\mathrm{EA}} \equiv \frac{\Delta\alpha^a}{\Delta\alpha^i} \propto \left(\mu_c^{1/4} + \mu_v^{1/4}\right)^{-4} \tag{3.30}$$

holds. This suggests that the EA polarization ratio, R_{EA}, is directly linked with carrier mobilities, although no one-to-one correlation between the R_{EA} ratio and either the electron mobility μ_c or hole mobility μ_v is well established. Finally, it should be remarked here that the condition, $W_v = (5/3)^{1/2}W_c(V_v/V_c)$, employed for the calculations resulting in Figs. 3.2 and 3.3 is identical with $\mu_c \simeq 3\mu_v$, which will also be shown to hold for vast classes of a-Si:H alloys so far as as-grown and/or annealed states of the materials are concerned.

3.1.3.2 Hall Effect Near the Mobility Edge

An alternative way to approach carrier mobilities would be to measure the Hall mobility. As is well known, however, most amorphous semiconductors exhibit "sign anomaly" which awaits theoretical interpretations before the method can be utilized as a meaningful tool for material characterization. Theories developed in the context of a hopping model can interpret the anomaly, but their validity in various classes of amorphous semiconductors is questionable, particularly for a-Si:H alloys. In terms of the mobility-edge model, no satisfactory explanation of the phenomenon has appeared. The failure is probably due to an unconscious neglect of the random interference of electron waves leading to the occurrence of a mobility edge.

We have attempted to give a quantitative description of the anomalous Hall effect near the mobility edge in accordance with a microscopic model in which interference effects are properly taken into account, as well as to decode the message from Hall measurements regarding carrier mobility [7]. New perturbative renormalization-group theory has been developed to tackle the problem. The Hall conductivity σ_H is then found to be

$$\sigma_H(\epsilon) = -\frac{B}{\pi^3}\frac{e^2}{\hbar\Lambda}\left(\frac{e}{\hbar}\Lambda^2\sqrt{\frac{\pi}{3}}u\right)\cdot\frac{u-1}{u^2}\left[u-1+\frac{5}{2}\frac{1}{u}-5\ln\left(\frac{u}{u-1}\right)\right] , \qquad (3.31)$$

where B is the magnetic field, and u denotes the energy measured from the band-edge, normalized by the width of the localized states at the band edge; so that $u = 1$ corresponds to the mobility edge. It is then straightforward to calculate the Hall mobility μ_H defined by σ_H/σ_B using Boltzman's statistics. The result is shown in Fig. 3.6, which illustrates how the Hall mobility of negatively charged carriers (electrons) behaves with the carrier mean free path. For $1 < \Lambda/a < 5$, the Hall mobility is found to be positive (anomalous) and increases with Λ/a, reaching about 0.1–0.15 cm^2 V^{-1} s^{-1}. The number appears to be about 1/100 of the conductivity mobility by comparing Fig. 3.4 and Fig. 3.5. Another striking feature seen from this figure is a reversal to normal Hall sign, which occurs at $7 < \Lambda/a < 14$, depending upon V/kT. The critical mean free path is in the range 16 Å$< \Lambda < 30$ Å. This kind of sign switch from anomalous to normal has been reported on microcrystalline silicon (mc-Si) with crystalline size greater than about 20 Å, implying the plausibility of the theory. Figure 3.6 permits us to evaluate the mean free path of the majority carrier from the measured Hall mobility. When both the EA polarization ratio R_{EA} and Hall mobility μ_H are measured on a single material, then mobilities of electrons and holes are separately determined through (3.24), (3.29), and Fig. 3.6, without any optional assumption on R^2.

3.1.3.3 Extended Model for Mobility Edge Transport

The discussions described above relies on the simple mobility-edge model, in which any factors that weaken the localization of the electron wave function

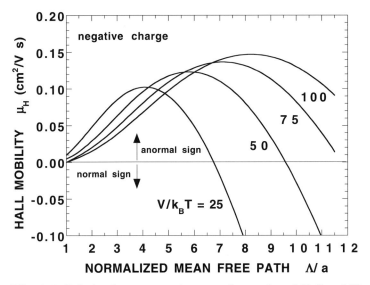

Fig. 3.6. Relation between carrier mean free path and Hall mobility for negatively charged carrier

are ignored, i.e., a strong magnetic field, inelastic scattering, high-frequency electric field, and so on. What we have to examine here is the effect of inelastic scattering that might dominate the carrier transport in the high-temperature condition being encountered in the practical outdoor operation of hydrogenated amorphous silicon-based solar cells. It is, however, not straightforward to take the inelastic scattering effect into account in the quantum-statistical mechanics treatment, instead we follow the renormalization-group (scaling) equation for the frequency-dependent differential conductivity due to Kawabata [8]:

$$\left[1 + \left(\frac{1}{u} - 1 \right) \frac{\partial}{\partial \ln u} - \left(\frac{1}{u} + 2 \right) \frac{\partial}{\partial \ln s} \right] \hat{\sigma}(\epsilon, \omega) = 0 \ , \tag{3.32}$$

where s denotes the normalized complex frequency defined by $i\omega\tau$ with τ being the relaxation time $\hbar/2\Delta$ in the vicinity of the band-edge, and

$$\sigma(\epsilon, \omega) = \frac{1}{\pi^3} \frac{e^2}{\hbar \Lambda} \hat{\sigma}(\epsilon, \omega) \tag{3.33}$$

corresponding to (3.10). Equation (3.32) is derived from the original Kawabata's formulation through mathematically manipulating it to accommodate for the infinitely large disordered semiconductor system of our particular interest. The solution of (3.32) is achieved with the aid of the perturbative result:

$$\hat{\sigma}(\epsilon, \omega) \cdot \left(\hat{\sigma}(\epsilon, \omega) - (u - 1) \right)^2 = \frac{3\pi^2}{4} u \left(i\omega\tau + \left(\frac{\Lambda}{L_i} \right)^2 \right) \ . \tag{3.34}$$

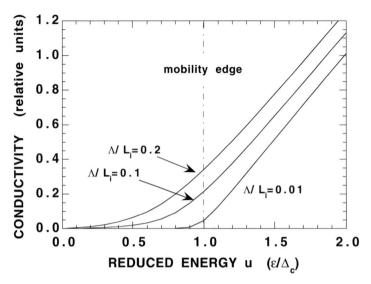

Fig. 3.7. band-edge conductivity spectra plotted for various normalized inelastic scattering lengths

where the effect of the inelastic scattering is included phenomenologically in terms of the inelastic scattering length L_i, being proportional to T^{-p} ($p \cong$ 1–2).

Equation (3.34) gives a new insight into the "ac complex conductivity" from the mobility-edge carrier-transport scheme, whereas our focus in this section is placed on the zero-frequency case, so that we put $\omega = 0$ in (3.34). The conductivity spectrum near the mobility edge is calculated on the basis of (3.34), and is shown in Fig. 3.7, implying that the conductivity spectrum shifts downward in energy with enhanced tails as the inelastic scattering length L_i is decreased, that is, for higher temperatures. The trend is interpreted as a physical consequence of the shortened coherence length of the electron wave function that weakens the localization. The inelastic scattering seems to make a large impact on the conductivity $\sigma(T)$ and thermopower $S(T)$ as well as on the macroscopic total charge carrier mobility $\mu(T)$. Equation (3.34) offers us a new possibility to discuss the temperature dependence of these transport-related quantities including the Q-function, $(e/k_B)S(T) + \ln \sigma(T)$, from the microscopic point of view; the conventional interpretation has been built on the macroscopic potential fluctuation model [9].

We will here give one example of such augments. Figure 3.8 shows the temperature dependence of the Q-function calculated in accordance with the solution of (3.34). It is found from this figure that the Q-function would be expressed approximately by

$$Q(T) = 9.3 - \ln\left(\Lambda \, [\text{nm}]\right) - 0.64\frac{\Delta_c}{k_B T} \,, \tag{3.35}$$

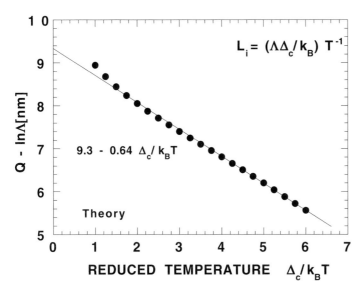

Fig. 3.8. Theoretical simulation for the temperature dependence of the Q-function

which reproduces the general behavior of the Q-function experimentally observed on a-Si:H, although a specific condition is assumed for the magnitude of the inelastic scattering length L_i as indicated in the figure. Comparing (3.35) with experimental results on phosphorus-doped n-type a-Si:H implies that the mean free path Λ of electrons is around 1 nm, and the width of the band-edge localized states Δ_C lies in the range of 100 meV to 250 meV depending on the doping concentration. We do not want to dwell on more details because these may be beyond the scope of this chapter. Finally, it should be remarked that the magnitude of the mobility, which we are interested in, is well represented by (3.39) so far as the room-temperature condition is concerned for a-Si:H alloys. For this reason, we will confine ourselves to the zero-temperature description of the transport parameters below, which, however, may not result in any fatal misleading of the physics.

3.2 Mobility and Band-Edge Parameters in Amorphous Silicon Alloys

3.2.1 Evaluation Procedure

Details of the experimental procedure are reported in our previous papers [3], so that we focus on the experimental results of our primary concern in this section. Before starting a description on specific results, it would be instructive to note that the dipole matrix element squared R^2 appears to slowly increase from $R^2 \cong 10$ Å^2 ($\Lambda/a \to \infty$) upon increased structural disorder, however remains in a rather narrow range $(10.65 \pm 3\%)$ Å^2 for the practical range of the mean free path, $1 \leq \Lambda/a \leq 5$, so that $R^2 = 10.65$Å^2 is employed for a-Si:H alloy materials for which Hall measurements cannot be performed. Combining (3.34) and (3.35) then leads to $\mu_c \simeq 15.3$ cm^2 V^{-1} s^{-1} ($\Lambda_c/a \simeq 5$) and $\mu_v \simeq 5.1$ cm^2 V^{-1} s^{-1} ($\Lambda_v/a \simeq 3$), which do not contradict those deduced from various related measurements, and might be regarded as the best numbers for current solar-cell grade undoped a-Si:H.

3.2.2 Carrier Mobility in Amorphous Silicon Alloys

3.2.2.1 Effects of Alloying and Growth Temperature

The EA polarization ratio R_{EA} measured on undoped a-SiGe:H, a-Si:H and a-SiC:H alloys is summarized in Fig. 3.9 as a function of the Tauc optical gap E_o. It should be stressed that the bandgap dependence itself has no physical significance in the sense that the material property can never be determined by the Tauc optical gap. Immediately read from this figure is that the lowest R_{EA} occurs for a-Si:H and it continuously increases upon alloying, which is probably due to the effect of alloy-induced disorder.

The data is translated into carrier mean free path and then mobility according to the procedure described above. The result is displayed in Fig. 3.10, where electron and hole mobilities are plotted against Tauc gap. The largest mean free paths ($\Lambda_c = 11.8$ Å, $\Lambda_v = 7.1$ Å) and mobilities ($\mu_c \simeq 15.3$ cm^2 V^{-1} s^{-1}, $\mu_v \simeq 5.1$ cm^2 V^{-1} s^{-1}) appear for a-Si:H prepared under conditions optimized for solar-cell application.

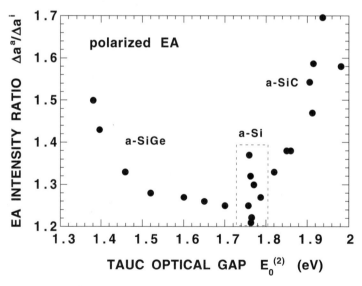

Fig. 3.9. PEA intensity ratio plotted against Tauc optical gap for a-Si:H alloys

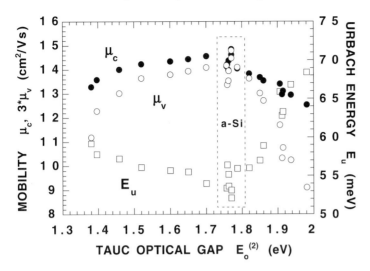

Fig. 3.10. Summary of carrier mobilities and Urbach tail energy for a-Si:H alloys with various Tauc optical gaps

The mobility tends to get reduced upon alloying, particularly in the case of C-alloying. Alloying with Ge does not lead to any severe effect on the electron mobility for concentrations upto about 35 at. % (E_o=1.5 eV), beyond which it sharply decreases down to about 85% of the a-Si:H value. The same quantitative tendency holds for the hole mobility, whereas its value is roughly about 1/3 of the electron mobility. On the other hand, a more pronounced reduction of mobilities is found for C-alloying. More than a 10% decrease results from a carbon concentration of only 12 at. % ($E_o = 1.98$ eV) for both electron and hole mobilities.

It is widely accepted that the lowest defect density is achieved for a specific growth (substrate, deposition) temperature, which naturally depends on other preparation conditions such as deposition rate. A fundamental question arises how about mobilities. The answer is demonstrated in Fig. 3.11, in which carrier mobilities, Urbach energy E_u and photosensitivity σ_{ph}/σ_d are plotted against the growth temperature. The growth rate was kept at 2.5 Ås^{-1}, and the hydrogen content varies in the range from 13.5 at. % (250°C) to 17.5 at. % (100°C). It is found that carrier mobility continuously increases with increasing T_d up to about 200°C, beyond which it decreases. Other physical quantities also take extremes near around this temperature, which seems to be fortuitous for us working in solar-cell applications. The magnitude of room-temperature mobility is essentially governed by structural static disorder involved in the materials. The structural disorder is likely to be determined by the extent of structural relaxation that occurred during the growth process in the low-temperature deposition case, while for the high-temperature deposition case it would be dominated by the thermal disorder is to be frozen-in during the

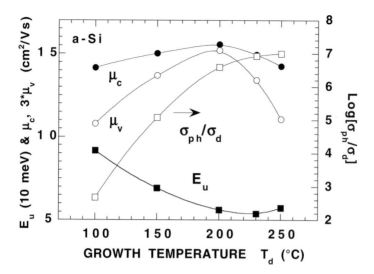

Fig. 3.11. Growth temperature dependence of various physical quantities in undoped a-Si:H

cooling process after deposition. These two competing mechanisms would operate to yield a mobility maximum at a particular growth temperature. It should again be noted that the optimum growth temperature "200°C" found in the present experiment can never be a universal quantity characteristic of plasma CVD a-Si:H, but it should be largely dependent on some details of deposition conditions, i.e., deposition rate, flux and energy of plasma species including ions impinging on a growing surface, cooling rate after deposition, and so on.

3.2.2.2 Doping Effect on Mobility

A reduction of TOF (total charge) mobility is observed in doped a-Si:H. A significant deviation of the activation energies for the conductivity and thermopower is also found. These are conventionally believed to be due to the long-rage electrical potential fluctuation associated with randomly distributed charged impurities and/or defects. If this idea is correct, then the EA polarization ratio R_{EA} should not be significantly affected by doping since it is essentially insensitive to the long-range potential fluctuation. What we have observed is, however, an increase in R_{EA} upon phosphorus doping, to the extent, which have never be experienced on undoped a-Si:H alloys, as demonstrated in Fig. 3.12. The electron Hall mobility decreases from about 0.11 cm^2 V^{-1} s^{-1} to 0.04 cm^2 V^{-1} s^{-1} with doping upto the level compatible with standard solar cell n-layer.

The experimental data in Fig. 3.12 are translated into the behavior of free carrier mobilities in accordance with the procedure described in the

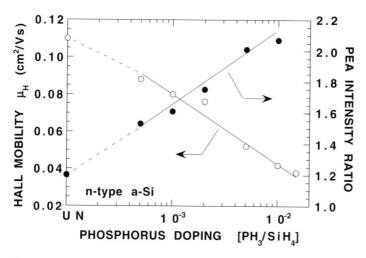

Fig. 3.12. Phosphorus doping dependence of the Hall mobility and PEA intensity ratio in n-type a-Si:H

previous section. As found in Fig. 3.13, the electron mobility decreases to about one-third of that in the undoped case, while the hole mobility remains almost unchanged. If the long-range potential fluctuation plays a critical role in determining the transport property, then both the electron and hole mobilities should be equally reduced by the incorporation of

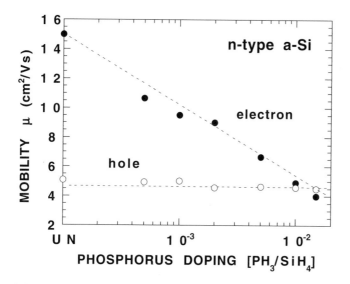

Fig. 3.13. Phosphorus-doping effect on the electron and hole mobilities in n-type a-Si:H

charged impurities and/or defects, which is, however, in contradiction with our experimental observation. Furthermore, although not systematic data, we have found a qualitatively identical behavior of mobilities for boron-doped materials; a decrease in the electron mobility, but no significant reduction in the hole mobility. These findings seem to imply that the doping-induced change in mobilities should not be attributed to classical charge effects, but the incorporation of impurity atoms itself may be of central importance.

3.2.2.3 Network Disorder and Carrier Mobility

Both the mobility and defect density are tightly linked with the disorder in the amorphous network structure; the latter basically reflects the short range order in the chemical bonding while the former depends on various classes of disorder from short-range (bond length, bond angle) to intermediate range (dihedral angle and topological issues including ring statistics). Various kinds of disorder can not be independent of each other, so that there may be some correlation between the mobility and defect density. The result in Fig. 3.11 indicates such an indirect correlation. In general, however, these two physical quantities are believed to be independent, and can be optimized on a single material through a fine control of the material growth process.

We now briefly discuss the physical meaning of the observed change in carrier mobilities in association with the network disorder. Incorporation of group IV elements like Ge and C leads to an almost parallel decrease in both the electron and hole mobility. On the other hand, doping with group III and V elements gives rise to a dramatical reduction of the electron mobility, while virtually retaining the hole mobility. From the theoretical viewpoint, it has been suggested that the fluctuation in the bond angle and dihedral angle predominantly impacts on the top of the valence band, while the bottom of the conduction band is quite sensitive to the topological disorder of the network; for example, an existence of an odd-membered ring, position of impurity atoms and network configurations. The majority of group III and V impurities are incorporated with three-fold coordination in the tetrahedrally bonded amorphous-silicon network, unless small doping conditions apply. The incorporation of three-fold coordinated impurity atoms induces an increase in network topological disorder, and brings about a particular affect on the conduction band-edge; electron mobility. The growth temperature and alloying will produce influences on the short-range chemical disorder and the intermediate/long-range network disorder as well, so that both the electron and hole mobilities are therefore subject to them to almost the same extent.

3.3 Photoinduced Structural Change

3.3.1 Photoinduced Changes in Electronic Properties

The photoinduced degradation of the electronic properties in a-Si:H, termed the Staebler–Wronski effect, has been intensively studied from both the basic physics and engineering points of view. Most of the previous studies devoted on this subject placed the focus on a photoinduced dangling-bond defect with an unpaired spin at $g = 2.0055$ that is believed to be a dominant nonradiative recombination center. Recently, several pieces of experimental evidences have appeared implying the photoinduced structural change occurred in addition to the photoinduced defect creation. If the photoinduced structural change were identified in physical quantities directly reflecting the overall structural disorder and correlated with the photoinduced defect creation, then a new possibility would emerge to understand the mechanism of the Staebler–Wronski effect, as well as to find a way to eliminate the photoinduced degradation. We have approached the problem by means of the polarized electroabsorption method, and observed a reversible photoinduced structural change taking place besides the changes in the photoconductivity as well as the dangling-bond density, both of which exhibit quite different time-evolution characteristics under light exposure and thermal annealing processes. In this section, we argue the critical role of the photoinduced structural change played in the Staebler–Wronski effects [10].

3.3.2 Photoinduced Structural Change and Its Physical Implications

3.3.2.1 Photoinduced Change in the Network Disorder

Figure 3.3 displays the EA spectra, $\Delta\alpha^i$ and $\Delta\alpha^a$, for a device-quality undoped a-Si:H in the annealed state, while Fig. 3.14 shows those for the light-soaked state created by a He-Ne laser beam exposure (632.8 nm, 27 mW cm^{-2}) of one hour. Solid lines are theoretical lineshapes fitted to each EA spectrum, from which the EA intensity ratio $\Delta\alpha^a/\Delta\alpha^i$, R_{EA}, is read out. It is clearly found that the R_{EA} increases from 1.2 to 2.0 upon light soaking, which is accompanied by a small but detectable red shift and the spectral broadening of both the EA spectra. The experimental observation definitely indicates that the degree of structural disorder increases upon light soaking.

The photoinduced degradation is known to be reversible upon light exposure and subsequent thermal annealing. If the change in the R_{EA} is a physical phenomenon closely correlated with the photoinduced degradation, then it should also be a reversible process. The experimental confirmation is given in Fig. 3.15, which demonstrates how the ratio R_{EA} behaves under the sequence of light exposure and thermal annealing. Upon light exposure, the ratio R_{EA} grows at the rate inversely proportional to the illumination intensity, and restores its original magnitude upon sufficient thermal annealing (170°C, 8 h).

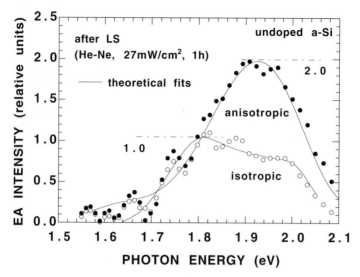

Fig. 3.14. Polarized electroabsorption spectra on undoped a-Si:H in the light-soaked state. Comparison is made with Fig. 3.3

Highly contrasted with the known behaviors of the metastable photoconductivity and dangling bond defects, the saturation level of the ratio R_{EA} appears to be independent of the illumination intensity, being in the range of 7.5–27

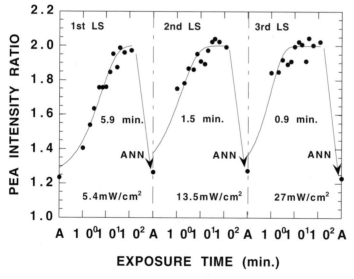

Fig. 3.15. Change in the PEA intensity ratio during consecutive light exposure and annealing cycles, with different illumination light intensities in undoped a-Si:H

mW cm^{-2}. The observation seems to be an indication that the saturation of the structural disorder is not determined by a balance between photoinduced enhancement and thermal relaxation, unlike the metastable steady states for defect-related physical quantities.

Although not shown here, it is found that the photoconductivity remains virtually constant at the beginning of light exposure but leads to decrease just after the saturation of the ratio R_{EA} occurs. The decay of the photo-conductivity is ascribed to the creation of metastable dangling bonds, which continues before the steady-state is reached; i.e., it takes much more expo-sure time than the saturation of R_{EA}. If an occurrence of the photoinduced structural change is true, then it should be manifested in more macroscopic properties like volume or mass density as in chalcogenide glasses. Motivated by this expectation, we have tried to pursue them on a-Si:H films by means of the optical bending-beam and flotation methods. Apart from quantitative details, quite similar reversible changes have been found for the internal stress and mass density as for R_{EA}, which may provide us with good evidence for the photoinduced structural change that occurred in the whole material.

These findings lead us to speculate that the photoinduced increase in the structural disorder may be a precursor process for the dangling-bond creation. Upon light exposure, nonequilibrium electrons and holes are generated and recombine nonradiatively through either band tails or native defects. The dis-sipated energy would allow groups of atoms to establish a local configuration of a higher energy, yielding a strain field spread over the whole network struc-ture. As a result, the overall structural disorder is increased including the fluctuation of the bond angle from its normal value at about 109°C, which manifests itself in the changes of the ratio R_{EA} as well as the internal stress. In the strained network structure of higher energy, dangling bond defects oc-cur via a large-scale atomic reconfiguration involving hundreds of Si atoms. In other words, the increase in structural disorder is followed by the dangling-bond creation, giving a possible explanation for the different temporal behav-iors of the ratio R_{EA} (network disorder) and photoconductivity (metastable dangling-bond density). More detailed discussion is given in our recent paper [11] and references therein.

3.3.2.2 Impact on Carrier Mobilities

It is logical to consider that the photoinduced structural change makes an appreciable impact on the carrier mobility, as implied by (3.30). Figure 3.16 demonstrates the temporal behavior of the R_{EA} and MPC-derived electron mobility μ_c in the light-exposure process. At a first glance, both the physical quantities exhibit a complementary change, that is, the R_{EA} increases and μ_c decreases upon light exposure. However, from a quantitative viewpoint, the photoinduced decrease in μ_c is too trivial if compared with the change shown in Fig. 3.4, which indicates $\mu_c \propto R_{EA}^{-1} \propto E_u^{-5.2}$ for a series of a-Si:H grown with different temperatures. As is well recognized, the Urbach energy E_u is

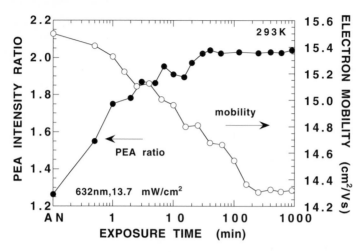

Fig. 3.16. Changes in the PEA intensity ratio and electron mobility upon light exposure for undoped a-Si:H

not influenced by light exposure, also being in contradiction with the behavior observed in Fig. 3.4.

The difficulty is easily solved if we accept that the photoinduced change in the network structure enhances the valence-band-related disorder selectively. A result of simulation, assuming $E_{uc} = 22$ meV $\to 23$ meV and $E_{uv} = 46$ meV $\to 87$ meV upon light exposure, is illustrated in Fig. 3.17. Note that the initial condition, $E_{uc}=22$ meV and $E_{uv}=46$ meV, fits well with the magni-

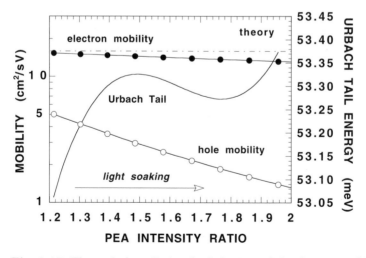

Fig. 3.17. Theoretical prediction for bahaviors of the electron and hole mobilities, Urbach tail energy and PEA intensity ratio in the light-soaking process

tudes of both the electron and hole mobilities, the PEA ratio R_{EA} and the Urbach energy E_u for device quality undoped a-Si:H material as discussed in previous sections. For μ_c and E_u, the simulation well reproduces experimental observations. There is a surprising prediction of a significant decrease in the hole mobility μ_v, since the photovoltaic performance of a-Si:H-based solar cells is mostly governed by the hole transport. Upon light exposure, only the dihedral angle as well as bond angle are to be subject to appreciable fluctuations that then would result in the increase in the valence-band disorder, rather than that in the conduction-band disorder. In this sense, the simulation result, Fig. 3.17, might not be so unrealistic. Extensive studies are clearly needed on this point from both the experimental and theoretical viewpoints not only to explain the physics of photoinduced changes but also to establish the best solar-cell design for a higher stability.

3.4 Concluding Remarks

This chapter has confined itself to a concise description of some specific topics concerning the electrical and optical properties of amorphous-silicon alloys. Not discussed here are the density distribution and nature of the localized states, as well as the dynamic communication processes between band carriers and these localized states that govern the transport property of photocreated nonequilibrium carriers, including the recombination process. The 20th century standard models for these issues are given in [3], which are believed to still work well if only the semiquantitative aspect of nonequilibrium carrier transport properties are concerned. However, it is clear that refined understandings of equilibrium and non-equilibrium carrier-transport processes are required for the establishment of high-efficiency and stable a-Si:H solar cells from the viewpoint of not only the material science but also the device technology.

Just a limited number of related articles are noted below, however, the reader is recommended to see references listed therein to gain more information.

References

1. Mott, N.F. (1987): Conduction in Non-Crystalline Materials, Oxford Science Publishing: Chap.3.
2. Economou, E.N., Soukoulis, C.M., Cohen, M.H., John, S. (1987): Disordered Semiconductors, ed. by Kastner, M.A. et al., Plenum Press: pp.681–701.
3. Okamoto, H. (1999): Amorphous Silicon, ed. by Tanaka, K., John Willey & Sons: Chap.4.
4. Okamoto, H., Hattori, K. (1996): Phenomenological Scaling of Optical Absorption in Amorphous Semiconductors, J. Non-Cryst. Solids 198&200: pp.124–127.

5. Cody, G.D. (1984): Hydrogenated Amorphous Silicon, Semiconductors and Semimetals vol. 21B, ed. by Pankove J.I., Academic Press: Chap.2.
6. Okamoto, H., Hattori, K. (1991): A Theoretical Consideration on the Electroabsorption Spectra of Amorphous Semiconductors, *J. Non-Cryst. Solids* 137&138: pp.627–630
7. Okamoto, H., Hattori, K. (1994): Hall Effect near the Mobility Edge, *J. Non-Cryst. Solids* 164&166: pp.445–448.
8. Okamoto, H. (1998): Electronic Properties of Amorphous Semiconductors (in Japanese) *Oyo-Butsuri* 67: pp.1419–1423.
9. Overhof, H., Thomas, P. (1989): Electronic Transport in Hydrogenated Amorphous Silicon, Springer Tracts in Modern Physics vol.114, Springer-Verlag: Chap.5.
10. Okamoto, H. (1997): Photoinduced Structural Changes in Hydrogenated Amorphous Silicon (in Japanese) *Oyo-Butsuri* 66: pp.1041–1046.
11. Okamoto, H. (1999): Photoinduced Structural Changes in Hydrogenated Amorphous Silicon, Tech. Dig. Int. PVSEC-11: pp.199–202.

Preparation and Properties
of Nanocrystalline Silicon

Michio Kondo and Akihisa Matsuda

The potential application of nanocrystal silicon much smaller than the carrier diffusion length of crystalline silicon to high-efficiency solar cells with over 10% efficiency has been unveiled. The advantage of nanocrystalline silicon over conventional polycrystalline silicon arises from the hydrogen passivation of electrically active defects, as well as contaminants, thanks to the low temperature fabrication process under a hydrogen atmosphere. The carrier recombination at the grain boundaries is significantly reduced by the hydrogen passivation and thereby an open-circuit voltage as high as 550 mV can be obtained for nanocrystalline silicon 10 nm in diameter.

The success of nanocrystalline silicon for solar cell-application has explored novel material science issues: (1) The growth mechanism of crystalline silicon at temperatures much lower than the melting temperature, (2) the role of hydrogen incorporated in the film and (3) the role of the grain boundary in electrical properties. The remarkable differences from the conventional polycrystalline silicon are related to hydrogen in the film, and the hydrogen is a consequence of the low-temperature processing.

In this chapter, we focus upon the role of hydrogen in the low-temperature growth process and the material properties of nanocrystalline silicon.

4.1 History of Nanocrystalline Silicon

Non-single-crystalline silicon is referred to by a variety of names depending on its crystallite size; nanocrystalline, microcrystalline, polycrystalline, multicrystalline silicon, if ordered by the crystallite size. In this chapter, these are roughly discriminated by its crystallite size, i.e., the "nanocrystalline silicon" implies a crystallite with a typical size of 10 nm.

Low-temperature synthesis of nanocrystalline silicon was discovered by Veprek and Marecek in 1968 [1]. A crystalline silicon source located at room temperature is etched out by atomic hydrogen generated by a remote plasma and the etching products are transferred to the heated substrate (originally the substrate temperature was around 600°C, and later a much lower temperature was employed). The reduction in the etching rate of silicon at higher temperatures [2] gives rise to the deposition on the substrate. In addition, the etching rate of amorphous silicon is higher than that of crystalline silicon. As a consequence, the chemical equilibrium between the crystalline silicon source and a deposited film on the substrate is established to form nanocrystalline silicon.

The chemical transport experiment is identical to the currently employed plasma CVD processes in the respect that the deposition precursors are silane-related radicals such as SiH_x and that atomic hydrogen plays a crucial role, because these species are also generated by the plasma from a monosilane-hydrogen gas mixture. The fabrication of nanocrystalline silicon using plasma CVD of silane started in 1979–1980. In 1979, the ECD group reported anomalously high conductivity of phosphorous doped amorphous silicon from SiF_4 source gas and ascribed this high conductivity to a more ideal Si network formed by the use of SiF_4 as compared to SiH_4 [3]. In 1980, they reported the presence of crystallites in the sample [4]. Japanese groups at approximately the same time reported the discovery of microcrystalline silicon synthesized by plasma CVD using high hydrogen dilution of silane. Usui and Kikuchi reported the heavily P-doped microcrystalline silicon [5]. Hamasaki et al. [6] and Matsuda et al. [7] independently reported the microcrystalline formation at low temperatures. The application to solar cells was attempted for the window layer due to its high conductivity and its low absorption coefficient in the visible region [8]. However, the microcrystalline silicon solar cell was not successful until the Neuchatel group reported 7% efficiency using gas purification and the VHF plasma technique [9, 10]. The predominant reasons that limit the efficiency were oxygen impurity and defect reduction. At present, an efficiency of more than 10% is obtained for single-junction cell [11] and the initial efficiency of 14.1% for a double-junction cell (1 cm^2) with an amorphous silicon top cell [12].

4.2 Preparation of Nanocrystalline Silicon

Plasma-enhanced chemical vapor deposition (PECVD) has been most widely used for preparing the nanocrystalline silicon. The advantage of the plasma process over the conventional CVD process is the capability of low-temperature synthesis of crystalline silicon due to the nonequilibrium process. In the conventional CVD processing, the high-temperature process needed for thermal dissociation limits the versatility of the substrate for solar cells, while in PECVD the film deposition is possible even at room temperature because the decomposition of the source gas, such as SiH_4 is caused by energetic electrons.

The spatial profile of the plasma potential is shown in Fig. 4.1. Since the electron has a much larger thermal velocity than the positive ion, the bulk

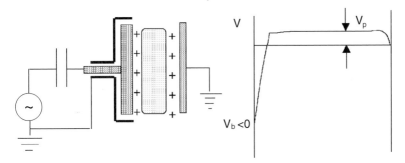

Fig. 4.1. Schematic view of the diode-type reactor for RF plasma CVD (left) and the spatial profile of the plasma potential between the electrodes. V_b and V_p are self-bias voltages of the cathode and plasma potential, respectively

Table 4.1. Principal emission lines from excited species in the hydrogen diluted silane plasma

Species	Emission wavelength (nm)	Threshold excitation energy (eV)
Si	288	10.53–11.5
SiH	414.23	10.33–10.5
H_α	656.28	12.7 (calculated)
H_β	486.13	12.09 (calculated)
H_2	602.13	14.0 (calculated)

plasma is required to have a positive potential in order to maintain the charge neutrality or to confine the electrons. The plasma sheath near the substrate accelerates the positive ions toward the substrate and gives rise to the ion bombardment. The plasma potential of the RF-plasma of silane is of the order of 10 eV, and the ion energy can be varied by applying the substrate bias.

The source gases are usually monosilane and hydrogen, and phosphine and diborane are used for doping. To improve the uniformity of the film thickness and material properties, the source gas is often supplied from the cathode electrode. The hydrogen dilution ratio, $R = [H_2]/[SiH_4]$ is a crucial parameter to determine the structure. As has been reported, the hydrogen dilution ratio has a threshold for crystal growth and by increasing the hydrogen dilution ratio the crystallinity is improved [13].

A useful characterization technique of the plasma is optical emission spectroscopy (OES). In the plasma, many molecules and radicals are in excited states due to collisions with the energetic electrons. The radiative transition from the excited states to the ground states or to the lower excited states can be detected optically and its photon energy is specific to each species. Table 4.1 shows typical emission lines for the plasma of silane/hydrogen mixture.

The excitation frequency is a dominating parameter of the plasma. Conventional RF plasma employs 13.56 MHz and a higher frequency above 60

MHz has been attempted recently [14, 9, 15, 16]. The frequency effect arises from the heavier mass of positive ions than electrons and also from the mechanism of energy gain of electrons in the plasma. With increasing drive frequency, the predominant mechanism of energy gain tends to be the stochastic heating (or the wave-riding effect) where the oscillating sheath boundary (that is the energy barrier for electrons) impacts on the electron to provide the kinetic energy, analogous to a ball hit by a racket. This is also termed as the "wave-riding effect". The average power gained by this mechanism, $< P_s >$, is given by the following expression,

$$< P_s >= \frac{1}{2} m_e n_e < v_e > w^2 \ , \tag{4.1}$$

where m_e, n_e and v_e are mass, density, and average velocity of electrons and w is the velocity of the sheath oscillation. If the applied RF voltage is kept constant, the sheath width is also constant and therefore the velocity, w, is proportional to the oscillating frequency, ω_o. This implies that the plasma density increases and that the plasma potential decreases with increasing plasma frequency.

4.3 Understanding Nanocrystalline Silicon Growth

The most remarkable aspect of the nanocrystalline growth is that the growth temperature of around 200–400°C is much lower than the melting temperature. On the basis of a simple energy diagram for amorphous and crystalline silicon as shown in Fig. 4.2, one can expect that the cause of the low-temperature formation of the crystalline phase is the extrinsic activation (A) or the reduction of energy barrier by some catalytic effect (or surfactant effect) (B). In this section, we discuss the growth mechanism of the low-temperature growth of nanocrystalline silicon in terms of the gas-phase reaction, as well as the surface reaction, particularly focusing upon the role of atomic hydrogen and ion bombardment.

Since the first synthesis of nanocrystalline silicon by the chemical transport method, atomic hydrogen has been considered to play an important role

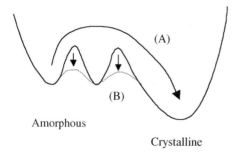

Amorphous

Crystalline

Fig. 4.2. Energy diagram for amorphous and crystalline phase of silicon

in the growth mechanism. In the chemical transport, the balance between etching by atomic hydrogen and deposition [1] is crucial. In the PECVD with a silane/hydrogen mixture, the correlation between the amount of atomic hydrogen and the crystallinity has been pointed out by an optical emission spectroscopy (OES) study [13].

Figure 4.3 shows a phase diagram depending on a wide variety of deposition conditions. In the low-RF power regime, where the ion bombardment is small, the crystalline phase appears in the high hydrogen dilution regime. This transition occurs sharply with increasing hydrogen dilution ratio, R. In the low dilution ratio regime, on the other hand, the structure is amorphous for low RF power, the crystalline phase appears in the high RF power region, and finally the amorphous phase again appears in the highest power region. The formation of microcrystalline silicon under high-density plasma with pure silane without hydrogen dilution has also been reported [17]–[19].

This phase diagram can be consistently interpreted in terms of atomic hydrogen density. The atomic hydrogen density is evaluated, for instance, by the emission intensity of $H\alpha$ lines ($\lambda = 633$ nm), and a good correlation with the appearance of the crystalline phase has been observed [13]. The

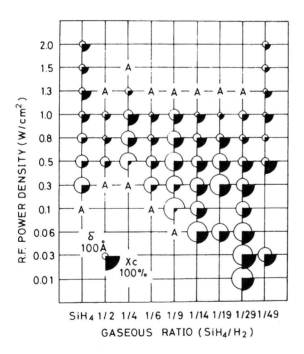

Fig. 4.3. Phase diagram of nanocrystalline silicon as a function of the gaseous ratio SiH_4/H_2 and RF power density. The radius of the circle and the closed symbol designates grain size and volume fraction, respectively (after [13])

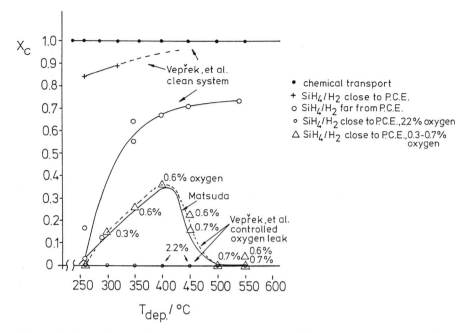

Fig. 4.4. Crystalline volume fraction as a function of deposition temperature for plasma-CVD and chemical-transport methods (after [20])

atomic hydrogen density is determined by the balance of the generation and annihilation rate in the gaseous phase. The most important reaction is the annihilation reaction by SiH_4, that is, $H + SiH_4 \rightarrow SiH_3 + H_2$, implying a very small amount of atomic hydrogen under pure silane or low hydrogen dilution conditions. Therefore, there are two possible ways to obtain sufficient amounts of atomic hydrogen; one is high hydrogen dilution. The other is silane-depleted conditions at high RF power because of the consumption of the silane molecule.

The temperature dependence of the crystalline volume fraction is shown in Fig. 4.4, and the deterioration in the crystallinity is seen in the high-temperature regime, $T > 400°C$ where the surface hydrogen starts to be thermally desorbed. This correlation shows the importance of the surface hydrogen coverage for crystal growth, as discussed later. However, Veprek et al. [20] have pointed out that the deterioration of the crystalline volume fraction is not observed under the ultra-high vacuum (UHV) conditions in their chemical transport method as shown in Fig. 4.4, while in the PECVD using a UHV system, the deterioration was confirmed at high-temperatures [21]. This discrepancy may be ascribed to the difference in atomic hydrogen density. With a high atomic hydrogen density in the chemical transport experiment, the thermal evolution of surface hydrogen can be recovered and a sufficient amount of surface coverage can be maintained even at high-temperatures. A similar high-temperature behavior is observed in the epitaxial growth [22].

The role of hydrogen has been extensively discussed in terms of (1) etching, (2) exothermic recombination of a pair of hydrogen atoms, (3) surface diffusion and (4) chemical annealing. The etching model [23] is based on the fact that the chemical-transport mechanism contains a critical balance between the etching and deposition on the substrate. In fact, the selective etching of amorphous silicon by atomic hydrogen does not create the crystalline structure. Therefore an additional mechanism to facilitate the crystal growth must be involved. (Note: In a personal discussion with Dr. Veprek, he clearly mentioned that he has never proposed a etching model as a cause of the microcrystalline silicon formation in spite of the fact that his paper is frequently cited as an etching model, see. p. 144, line 19–30 in [24])

The second model is based on the idea where the energy barrier from amorphous to crystalline structure in Fig. 4.2 is surmounted by the external energy. As a source of the external energy, atomic hydrogen recombination on the surface has been proposed [25]. This model can account for why higher hydrogen dilution is needed at lower deposition temperatures for improved crystallinity. It has also been proposed that the hydrogen abstraction by atomic hydrogen and the successive termination of the dangling bond by atomic hydrogen are both exothermic reactions; Si-H + H → Si- + H_2, Si- + H → Si-H, and these exothermic reactions facilitate the surface diffusion [26].

The surface-diffusion model is based on the fact that the main precursor, SiH_3 is weakly adsorbed and thereby very mobile on a hydrogen-covered surface [27]. In the growth of a-Si:H, a-SiGe:H, and a-SiC:H, this model successfully accounts for the surface smoothening effect and the improvement of film quality under the hydrogen-dilution condition [27, 28]. The disrupting role of a hydrogen-desorbed surface for nanocrystalline silicon growth has been demonstrated by the intermittent deposition [29], where the crystalline formation is disrupted for longer off-times because of the surface hydrogen desorption at a high temperature of 400°C. It is interesting that the role of hydrogen in the low-temperature MBE is disrupting, in contrast to the PECVD case [30]. This difference could be ascribed to the precursors, i.e, atomic Si in MBE and SiH_x in PECVD. The beneficial role of the surface coverage has also been reported in the case of a chlorine-terminated surface [31]. The role of chlorine or fluorine addition has been reported by several groups [32]–[34].

The subsurface reaction model termed as a chemical annealing model [35], claims the importance of the subsurface structural relaxation mediated by permeating atomic hydrogen on the basis of the fact that the atomic hydrogen exposure to the amorphous thin layer gives rise to the crystallization. The exothermic reaction of Si-H abstraction by atomic hydrogen in the subsurface region can facilitate the structural relaxation (phase transition). In this picture, the subsurface region with a thickness of a few nanometers does not establish its steady-state structure.

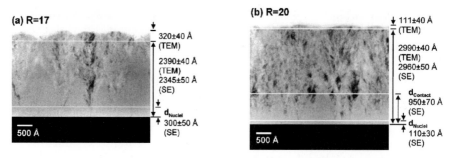

Fig. 4.5. XTEM views of nanocrystalline silicon on c-Si substrates covered by native oxide (after [43])

External energy can be provided by ion bombardment. The improved crystallinity in the low-temperature epitaxy using sputtering has been reported to be due to low energy ion flux [36]. Even at -180°C, microcrystalline silicon growth has been confirmed using reactive sputtering with H_2 [37]. On the other hand, the disrupting role of ions also has also been reported [16, 38, 39] in the PECVD processes from the substrate bias effects on the crystallinity. The positive ions in the plasma are accelerated in the sheath region as shown in Fig. 4.1. In the case of the usual RF plasma, the plasma potential is of the order of 10 eV. This energy is enough to destroy the Si-Si network.

The film structure is sensitive to the presence of impurities. Other impurities such as oxygen, water, and carbon can influence the crystal growth. The enlargement of the grain size in the temperature range above 350°C when using the UHV system [21], while in the low temperature regime the impurity effect on the crystal growth is barely observed, probably due to the complete surface coverage.

The nucleation of crystallites does not occur directly on the foreign substrate. As shown in Fig. 4.5, the crystal growth starts to occur after the formation of an amorphous incubation layer. The incubation thickness decreases with the hydrogen dilution ratio. The incubation layer is amorphous but has a rather different structure as compared to "normal" amorphous silicon, i.e., (1) higher compressive stress and (2) the presence of Si-H complex in the near-surface region. The compressive stress appears as a higher-frequency shift of the Raman spectrum. The usual amorphous silicon shows a TO phonon peak centered around 480 cm^{-1}, while the incubation layer shows a peak centered 485–490 cm^{-1} [40, 41]. The mechanical stress measurement also shows the increasing compressive stress an approaching the nanocrystalline phase boundary by increasing the hydrogen dilution ratio, R [42].

The presence of the Si-H complex in the incubation layer has been confirmed using an attenuated total reflection (ATR) method as shown in Fig. 4.6. The Si-H complex is observed at a frequency ($\sim$ 1940 cm^{-1}) lower than the usual Si-H stretching mode either in the bulk ($\sim$ 2000 cm^{-1}) or on the surface ($\sim$ 2090 cm^{-1}) [43]. This lower frequency shift implies the formation of the three center bond, Si-H-Si as shown in Fig. 4.6. The steady-state density of the complex increases with increasing hydrogen dilution ratio and growth temperature, and nucleation occurs when the density of this complex reaches the critical value [43].

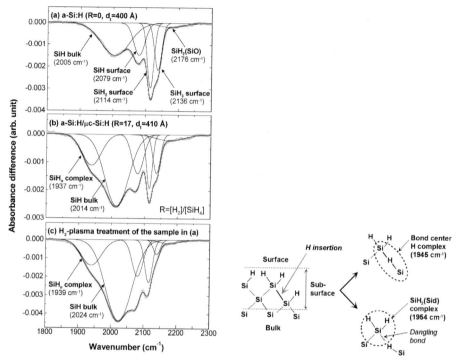

Fig. 4.6. ATR spectra for different hydrogen dilution and surface treatments (after [43])

4.4 High-Rate Growth of Nanocrystalline Silicon

It seems plausible that the low-temperature growth requires the critical balance of the formation of the regular structure and the removal of the irregular (disordered) structure [1]. Thus, it has been believed that the low growth rate is intrinsic to the μc-Si:H [44]. In fact, the low-temperature epitaxy has the maximum growth rate limited by the growth temperature [45]. From an industrial viewpoint, however, mass production for cost reduction is essential and the high growth rate is a crucial factor, and a growth rate above 50 Å s^{-1} is required assuming the thickness to be more than 2 μm. The growth rate is determined by the flux density of the film precursors such as SiH_x radicals, and the generation rate of radicals G_r is given by the following expression,

$$G_r = n_e v_e N_g \sigma_d \, , \qquad (4.2)$$

where n_e and v_e are the density and velocity of energetic electrons, respectively, with a kinetic energy higher than the dissociation threshold of the source gas molecule such as SiH_4. N_g and σ_d are, respectively, the density and the dissociation cross section of the source gas molecule.

A novel method, termed the high-pressure depletion (HPD) method [16, 46, 47], combines the high-pressure and high RF power conditions for a high growth rate of high-quality nanocrystalline silicon. The high gaseous pressure increases the gas density N_g and reduces the ion bombardment, while it decreases the atomic hydrogen density due to the recombination. The high RF power condition increases the electron density and suppress the hydrogen annihilation reaction, while it increases the ion bombardment. The combination of these two conditions enhances their advantages and compensates the drawbacks with each other. As a result, a growth rate of 15 Å s^{-1} with a reasonable crystallinity, and low ESR spin density of 2×10^{16} cm^{-3} are realized, and these values are reasonable for device application [46]. Further development has been achieved using a VHF plasma, and device-grade films with a defect density of 2.6×10^{16} cm^{-3} is obtained at 58 Å s^{-1} [16, 47, 48]. A similar high growth rate using a UHF plasma has been reported [17]. The suppression of the ion damage at high gaseous pressure is beneficial for suppressing the defect density, as shown in Fig. 4.7.

Nonplasma processes also have been extensively studied for nanocrystalline-silicon fabrication. The hotwire method (or catalytic CVD) method uti-

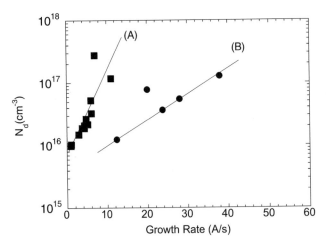

Fig. 4.7. Defect density measured by ESR as a function of the deposition rate under the conventional plasma conditions (A) and the high-pressure depletion conditions (B) (the vertical axis gives N_d in cm^{-3})

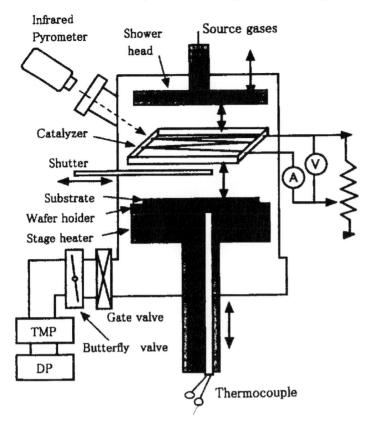

Fig. 4.8. Schematic of the deposition system using the HW-CVD method (after [49])

lizes the catalytic pyrolysis of source-gas molecules on the heated surface of specific metal such as tungsten [49]. The deposition apparatus is schematically shown in Fig. 4.8. The advantage of the catalytic process is efficient dissociation of hydrogen as well as silane. The fundamental mechanism of this process has recently made significant progress [50, 51], and material properties and the solar-cell performance have been improved [52]–[54] for not only a-Si:H but nanocrystalline-silicon solar cells. A hot-wire cell method that is similar to the cracking cell used in MBE has also been proposed and a growth rate as high as 2 nm s^{-1} without hydrogen dilution is obtained [55].

4.5 Structural Properties of Nanocrystalline Silicon

The nanocrystalline silicon commonly consists of small crystallites of the order of 10 nm and an amorphous silicon component. A crystallite > 10 nm shows nearly identical spectrum to the single-crystalline silicon centered at

520.5 cm^{-1} and its width mainly depends on the internal stress distribution. Amorphous silicon shows a broad and nearly symmetric lineshape centered at ~ 480 cm^{-1}. A simple calculation based on the phonon-dispersion relation and its size effect indicates the asymmetric lineshape and low-frequency shift of the peak position of the Raman spectrum for small crystallites < 10 nm as shown in Fig. 4.9 [56, 57].

It should be noted that a peak shift also occurs due to the mechanical stress in the film. For single-crystalline silicon, the Raman shift by stress has been extensively studied by Cerdeira et al. [58].

The volume fraction, F, of the crystalline phase in microcrystalline silicon is estimated by the following equation

$$F = I_c/(I_c + \gamma I_a) \ , \tag{4.3}$$

where the I_c, I_a, and γ are integrated intensity of amorphous and crystalline components, and scattering efficiency ratio, respectively. The value of γ has been estimated to be 0.88 [59]. Here, usually the contribution of the crystallites smaller than several nm are difficult to evaluate because the scattering efficiency is unknown.

The XRD measurements provide information on the lattice constant, crystallite size and orientation. The lattice constant is deviated from that of single-crystalline silicon due to strain, and the strain can be compared to the Raman peak shift. For instance, isotropic compressive stress induces a higher-frequency shift in the Raman peak and reduced lattice constant, while

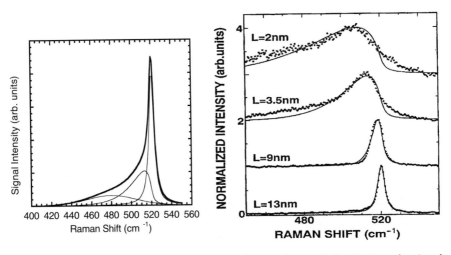

Fig. 4.9. Size dependence of Raman spectra (right, the vertical axis gives the signal intensity in arbitrary units) for size-controlled nanocrystalline silicon and a typical Raman spectrum (left) of mixed-phase sample prepared by PECVD. The dotted line in the left figure designates the amorphous, nanocrystallite ($\sim$ a few nm) and large crystallite (> 10 nm), components obtained from the simulation [57]

the biaxial strain due to such a thermal (compressive) stress results in the higher-frequency shift in the Raman peak and the expansion in the lattice constant because the XRD usually measures the lattice constant normal to the substrate.

The randomly oriented nanocrystalline silicon shows peak-intensity ratios 1:0:0.55:0.3 among (111), (220), and (311) diffraction peaks, respectively, whereas preferential orientation is found under various deposition conditions. The most commonly observed preference is the (220) orientation as shown in Fig. 4.10. The cross-sectional transmission electron micrograph (XTEM) shows a columnar structure in the film with the (220) preferential orientation. Other preferential orientation along (400) directions have been reported [34]. A difference in the etching rate among different orientation of crystallites may play an important role.

The grain size is commonly evaluated using Sherrer's formula. Strictly speaking, however, in the case of the anisotropic shape of the crystallite as shown in Fig. 4.9, this estimation can lead to a significant error. In fact, the TEM observation reveals the columnar shape with a diameter of about 20 nm and a longitudinal axis of 1 μm.

Nanocrystalline silicon prepared at low-temperatures contains a lot of hydrogen in the film, a typical density is around 3–5 at %. The hydrogen amount is rather lower than that of a-Si:H (10 at % typically) [60], and the hydrogen is considered to precipitate on the surface of grain boundaries. A simple calculation where columns with a diameter of 20 nm are packed closely indicates

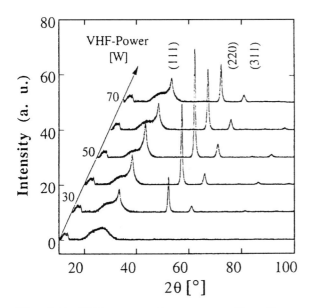

Fig. 4.10. XRD patterns for nanocrystalline silicon prepared at different plasma density (RF power) (after [10])

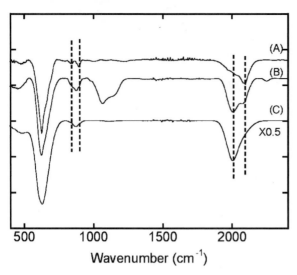

Fig. 4.11. R spectra for highly crystallized sample ($I_c/I_a = 8.5$; A), moderately crystallized sample ($I_c/I_a = 4.3$; B) and a-Si:H prepared at high growth rate (C). The four dashed lines designate 845, 890, 2000, and 2100 cm^{-1}, respectively. Sample (B) shows the absorption peak around 1000 cm^{-1} due to the post oxidization (the vertical axis gives the absorbance in arbitrary units)

the number of surface dangling bonds of the order of 1 at % of silicon, which is consistent with the observed amount of incorporated hydrogen. Figure 4.11 shows a typical IR absorption spectrum of highly crystallized nanocrystalline silicon, together with that of a-Si:H for comparison. The device-grade a-Si:H shows typical absorption peaks at 2000 and 630 cm^{-1}, while nanocrystalline silicon shows peaks at 2100 cm^{-1} and this is ascribed to the surface Si-H mode on the grain boundaries and/or the SiH$_2$ mode because the SiH$_2$ waging and scissors bending mode (845 and 890 cm^{-1}) are also observed. Poor crystallinity samples show both 2100 and 2000 cm^{-1} components. Since the μc-Si contains a high density of grain boundaries with a columnar structure, the grain boundaries tend to be oxidized during the air exposure. The post-oxidation affects the electrical conductivity as well as the defect density [61, 62].

4.6 Optical and Electrical Properties of Nanocrystalline Silicon

The optical absorption spectrum of nanocrystalline silicon measured by a constant photocurrent method (CPM) as shown in Fig. 4.12, compared with the single-crystalline silicon and hydrogenated-amorphous silicon. Since the thin-film sample does not have enough absorption, specific methods other than the usual transmission and reflection measurements are used. In the early stage of

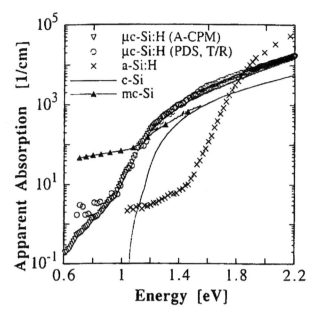

Fig. 4.12. Apparent absorption spectra for micro-(nano-)crystalline silicon (open symbols), single-crystalline silicon (solid line), multicrystalline silicon (closed symbol), and a-Si:H (x) (after [10])

the microcrystalline-silicon study, the enhancement of the optical absorption coefficient has been extensively argued because the absorption coefficient is crucial for crystalline silicon solar cells. A conclusion is that the enhancement is due to the optical scattering by a textured surface [63, 64]. The bandgap of nano (micro)crystalline silicon is measured to be ~ 1.1 eV at room temperature. This suggests the absence of the quantum size effect in the bandgap at a crystallite size of 10–20 nm.

As mentioned above, an impurity such as boron significantly influences the crystallinity and low-temperature deposition is needed for better crystallinity. The low-temperature deposition at around 140–180°C results in the formation of the B–H complex, as observed in single-crystalline silicon [65]. The B–H complex is observed at 1845 cm^{-1} by the IR absorption spectrum, and the electrical conductivity is quite low due to the passivation of boron acceptors. The post-deposition annealing above 200°C markedly increases the conductivity, and correspondingly, the absorption peak due to B–H complexes disappear in the IR spectrum, as reported in B-doped single-crystalline silicon [66].

The oxygen contamination can be simply eliminated by using a UHV system and/or using a gas-purification system, because the possible origin of the contamination is the source gas and/or the outgassing from the chamber walls [21]. Without using any purification system, therefore, a significant amount of oxygen of the order of 10^{19} cm^{-3} can be incorporated into the film. This oxygen is considered to form donor states in nanocrystalline silicon. As mentioned in Chap. 8, the suppression of this oxygen-related donor is a key element of the higher solar-cell efficiency. It has been reported that a very simple method to lower the deposition temperature effectively suppresses the oxygen donors, as shown in Fig. 4.13 [67].

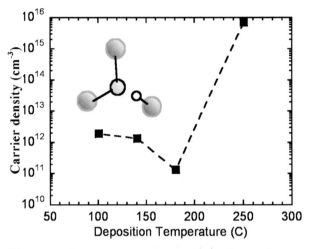

Fig. 4.13. Deposition-temperature dependence of carrier density ionized from oxygen related donors. The inset shows a model of the oxygen–hydrogen complex (after [67])

The donor concentration exceeds 10^{16} cm^{-3} at $T = 250°C$ and decreases by several orders of magnitude for $T < 180°C$. It was confirmed that the amount of oxygen is independent of the growth temperature. This implies that the oxygen has a different structure depending on the growth temperature. The fact that the post oxidation also results in an increase in the conductivity [62], and a similar electrically active oxygen is confirmed in polysilicon [68] indicates that the oxygen-related donors precipitate at grain boundaries. Although the structure responsible for this mechanism is not yet identified, a model analogous to oxygen donors in amorphous silicon [69] has been proposed, where the O atom forms the three-fold coordination. This picture explains the reason for the deactivation of the donor at low-temperatures. The excess atomic hydrogen at low-temperatures can be inserted into one of the three Si-O bonds to form Si-H and twofold coordinated oxygen as Si-O-Si. This is analogous to the B-H complex in c-Si as mentioned above. The passivation of the thermal donors in crystalline silicon has been discussed in terms of three-fold coordination [70].

Defects usually act as recombination centers in solar cells, and in the pin structure, the defects also act as space charge to screen the built-in field. Therefore the nature of defects in nanocrystalline silicon is of great importance. In a-Si:H, it is well known that the defect is a three-fold coordinated Si with a dangling bond. In nanocrystalline silicon, major defects are grain-boundary defects and these are passivated by hydrogen as mentioned above. The defect density for a device-grade sample is typically $< 2 \times 10^{16}$ cm^{-3}. This number is less than the density of the grains, assuming the grain size of ~ 10 nm, implying the number of defect per grains is less than unity. The neutral defect can be detected by electron spin resonance (ESR). The ESR signal (first derivative of the microwave absorption spectrum) for the highly [220] oriented sample is shown in Fig. 4.14. The g-value is nearly identical to that of amorphous silicon, while a more asymmetric lineshape is often observed. The grain-boundary defects are expected to show anisotropy with respect to the crystallite orientation. For the axially oriented sample along [220], angular-dependent ESR lines have been observed [71]. The anisotropies of the g-tensors are estimated to be $g_{\parallel} = 2.0023$, $g_{\perp} = 2.0078$, which are nearly the same as the Pb center [72].

The similarity of the transport properties between the amorphous and nanocrystalline silicon suggests the similarity of the density of states between amorphous and nanocrystalline silicon [73]. One of the cause of this similarity is the presence of the tail states. In amorphous silicon, tail states accompa-

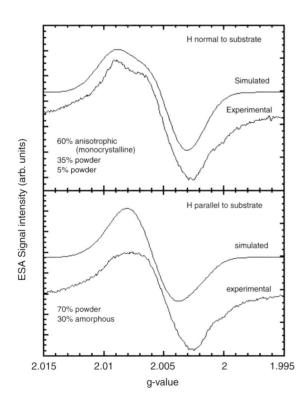

Fig. 4.14. Anisotropic ESR spectra of [220] oriented nanocrystalline silicon under perpendicular and parallel magnetic field to the substrate (after [40])

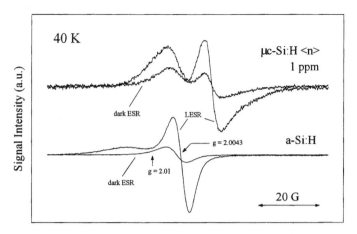

Fig. 4.15. LESR (nondoped sample) and ESR spectra of nanocrystalline silicon (upper traces) in comparison with those of a-Si:H (lower traces) (after [14])

nied with conduction and valence bands have been observed as light-induced ESR signals at g-values of 2.0044 and 2.010, respectively [74]. The conduction electron in crystalline silicon shows a g-value of 1.998, and a similar g-value is observed in amorphous silicon using a time resolved optically detected magnetic resonance (ODMR) [75].

In nanocrystalline silicon, a phosphorous-doped sample and an intrinsic sample illuminated by bandgap light show an identical resonance line at a g-value of 1.998 besides the dangling bond signal [14, 40, 76, 77, 78] as shown in Fig. 4.15. The origin of these lines is now ascribed to the conduction band tail states [40, 76]. The tail states compensate the donors and pinning of the Fermi level may occur, and the ionized donors act as scattering centers and as trapping centers for carriers. Although the origin of the tail states is not clear, random strain and a surface contribution are possible candidates and further study will be needed.

In summary, the development of the low-temperature processing of nanocrystalline silicon and the progress in the understanding of their material properties promotes the application of this material to solar cells. A more detailed discussion for solar cell processing will appear in Chap. 8.

Acknowledgments

The authors would like to acknowledge fruitful collaboration with Drs. Y. Nasuno, H. Yamamoto, S. Suzuki, T. Wada, H. Mase, T. Yamamoto, M. Fukawa, L. Guo, K. Saitoh, and H. Fujiwara. We are also indebted to Mrs. Y. Ohtake for the preparation of the manuscript.

References

1. Veprek, S., and Marecek, V. (1968): *Solid State Electron.*, 11 p.683.
2. Webb, A.P., and Veprek, S. (1979): *Chem. Phys. Lett.*, 62 p.173.
3. Madan, A., Ovshinsky, S.R., and Benn, E. (1979): *Phil Mag.* B, 40 p.259.
4. Tsu, R., Izu, M., Ovshinsky, S.R., and Pollak, F.H. (1980): *Solid State Commun.*, 36 p.817.
5. Usui, S., and Kikuchi, M. (1979): *J. Non-Cryst. Solids*, 34 p.1.
6. Hamasaki, T., Kurata, H., Hirose, M., and Osaka, Y. (1980): *Appl. Phys. Lett.*, 37 p.1084.
7. Matsuda, A., et al. (1980): Jpn. J. *Appl. Phys.* 20 p.183.
8. Uchida, Y., Ichimura, T., Ueno, M., and Haruki, H. (1982): *Jpn. J. Appl. Phys.*, 21 p.586.
9. Meier, J., et al. (1996): *Proc. Mater. Res. Soc. Symp.* 420 p.3.
10. Keppner, H., Meier, J., Torres, P., Fischer, D., and Shah, A. (1999): *Appl. Phys.* A, 69 p.169 and references therein.
11. Yamamoto, K., Yoshimi, M., Tawada, Y., Okamoto, K., Nakajima, A., and Igari, S. (1999): *Appl. Phys.* A, 69 p.179 and references therein.
12. Yamamoto, K. (2001): Tech. Digest of PVSEC 12, Jeju, Korea, p.547.
13. Matsuda, A. (1983): *J. Non-Cryst. Solids*, 59 & 60 p.767.
14. Finger, F., et al. (1994): *Appl. Phys. Lett.*, 65 p.2588.
15. Saitoh, K., et al. (1998): Proc. 2nd WCPEC, Vienna.
16. Kondo, M., et al. (2000): *J. Non-Cryst Solids*, 266–269 p.84.
17. Shirai, H., Sakuma, Y., Moriya, Y., Fukai, C., and Ueyama, H. (1999): *Jpn. J. Appl. Phys.*, 38 p.6629.
18. Scheib, M., Schroeder, B., and Oechsner, H. (1996): *J. Non-Cryst. Solids*, 198–200 p.895.
19. Endo, K., Isomura, M., Taguchi, M., Tarui, H., and Kiyama, S. *Solar Energy Mater. Solar Cells*, 66 in press.
20. Veprek, S., Sarott, F.A., and Ruückschloss, M. (1991): *J. Non-Cryst Solids*, 137&138, 733.
21. Kamei, T., et al. (1998): *Proc. Mater. Res. Soc. Symp.*, 507 p.867.
22. Kitagawa, T., Kondo, M., and Matsuda, A. (2000): *Appl. Surf. Sci.*, 159-160 p.30.
23. Tsai, C.C., Anderson G.B., and Thompson, R. (1990): *Mater. Res. Soc. Symp. Proc.* 192 p.475.
24. Veprek, S., Iqbal, Z., and Sarott, F.A. (1982): *Phil. Mag.*, B, 45 p.137.
25. Iqbal, Z., and Veprek, S. (1982): *J. Phys.* C, 15, 377.
26. Matsuda, A. (1996): *Thin Solid Films*, 337 p.1.
27. Matsuda, A., and Tanaka, K. (1987): *J. Non-Cryst Solids*, 97&98 p.1367.
28. Ganguly, G., and Matsuda, A. (1993): *Phys. Rev.* B, 47 p.3661.
29. Nomoto, K., Urano, Y., Guizot, J., Ganguly, G., and Matsuda, A. (1990): *Jpn. J. Appl. Phys.*, 29 p.1372.
30. Eaglesham, D.J. (1995): *J. Appl. Phys.*, 77 p.3597.
31. Guo, L., Toyoshima, Y., Kondo, M., and Matsuda, A. (1999): *Appl. Phys. Lett.*, 75 p.3515.
32. Platz, R., and Wagner, S. (1998): *Appl. Phys. Lett.*, 73 p 236.

33. Shirai, H., Arai, T., and Ueyama, H. (1998): *Jpn. J. Appl. Phys.*, 37 p.1078.
34. Kamiya, T., Nakahata, K., Ro, K., Fortman, C.M., and Shimizu, I. (1999): *Jpn. J. Appl. Phys.* 38 p.5750.
35. Akasaka, T., and Shimizu, I. (1995): *Appl. Phys. Lett.*, 66 p.3441.
36. Shindoh, W., and Ohmi, T. (1996): *J. Appl. Phys.*, 79 p.2347.
37. Hashimoto, S., Miyasato, T., and Hiraki, A. (1983): *Jpn. J. Appl. Phys.*, 2 L748.
38. Matsuda, A., Kumagai, K., and Tanaka, K. (1983): *Jpn. J. Appl. Phys.*, 22 p.34.
39. Konuma, M., Curtins, H., Sarott, F.A., and Veprek, S. (1987): *Phil. Mag. B*, 55 p.377.
40. Kondo, M., Toyoshima, Y., Ikuta, K., and Matsuda, A. (1996): *J. Appl. Phys.*, 80 p.6061.
41. Guha, S., Yang, J., Williamson, D.L., Lubianiker, Y., Cohen, J.D., and Mahan, A.H. (1999): *Appl. Phys. Lett.*, 74 p.1860.
42. Gotoh., T., Nonomura, S., Nishio, M., Nitta, S., Kondo, M., and Matsuda, A. (1998): *Appl. Phys. Lett.*, 72 p.2978.
43. Fujiwara, H., Kondo, M., and Matsuda, A. (2002): *Surf. Sci.*, 497 p.333.
44. Veprek, S., Glatz, F., and Konwitischny, R. (1991): *J. Non-Cryst. Solids*, 137–138 p.779.
45. Jorke, H., Herzog, J., and Kibbel, H. (1989): *Phys. Rev. B.*, 40 p.2005.
46. Guo, L., Kondo, M., Fukawa, M., Saitoh, K., and Matsuda, A. (1998): Jpn. *J. Appl. Phys.*, 37 p.1116.
47. Fukawa, M. et al. (2001): *Solar Energy Mater. Solar Cells*, 66 p.217.
48. Suzuki, S., Kondo, M., and Matsuda, A. (2001): Tech. Digest of PVSEC 12, Jeju, Korea, p.559.
49. Matsumura, H. (1998): *Jpn. J. Appl. Phys.* 37, p.3175, and references therein.
50. Tange, S., Inoue, K., Tonokura, K., and Koshi, M. (2001): *Thin Solid Films*, 395 p.42.
51. Nozaki, Y., et al. (2001): *Thin Solid Films*, 395 p.47.
52. Mahan, H. (2001) *Thin Solid Films*, 395 p. 12.
53. Schroder, B., Weber, U., Seitz, H., Ledermann, A., Mukherjee, C. (2001): *Thin Solid Films*, 395 p.298 .
54. Klein, S., Finger, F., Carius. R., Kluth, O., Neto, L., Wagner, H., and Stutzmann, M. (2002): Proc. EU-PVSEC Munich.
55. Ichikawa, N., Takeshita, J., Yamada, A., and Konagai, M. (1999): *Jpn. J. Appl. Phys.* 38 p.L24.
56. Campbell, I.H., and Fauchet, P.M. (1986): *Solid State Commun.*, 58 p.739.
57. Kanemitsu, Y., Uto, H., Masumoto, Y., Matsumoto, T., and Mimura, H. (1993): *Phys. Rev. B*, 48 p.2827.
58. Cerdeira, F., Buchenauer, C.J., Pollak, F.H., and Cardona, M. (1972): *Phys. Rev. B*, 5 p.580.
59. Tsu, R., Gonzalez-Hernandez, J., Chao, S.S., Lee, S.C., and Tanaka, K. (1982): *Appl. Phys. Lett.*, 40 p.534.
60. Brodsky, M.H., Cardona, M., and Cuomo, J.J. (1977): *Phys. Rev. B*, 16 p.3556.
61. Kondo, M., Nishimiya, T., Saitoh, K., and Matsuda, A. (1998): *J. Non-Cryst. Solids*, 227 p.890.
62. Goerlitzer, M., Beck, N., Torres, P., Kroll, U., Keppner, H., Meier, J., Koehler, J., Wyrsch, N., and Shah, A. (1998): *J. Non-Cryst. Solids*, 227–230 p.996.

63. Meier, J., et al. (1994): *Appl. Phys.* Lett., 65 p.860.
64. Vanecek, M., Poruba, A., Remes, Z., Beck, N., and Nesladek, M. (1998): *J. Non-Cryst. Solids*, 227–230 p.967.
65. Pankov, J.I., Zanzucchi, P.J., Magee, C.W., and Locpvsky, G. (1985): *Appl. Phys. Lett.*, 46 p.421.
66. Stavola, M., et al. (1988): *Phys. Rev.* B, 37 p.8313.
67. Nasuno, Y., Kondo, M., and Matsuda, A. (2001): *Appl. Phys. Lett.*, 78 p.2330.
68. Werner, J.H., Dassow, R., Rinke, T.J., Kohler, J.R., and Bergmann, R.B. (2001): *Thin Solid Films*, 383 p.95.
69. Morimoto, A., Matsumoto, M., Yoshita, M., Kumeda, M., and Shimizu, T. (1991): *Appl. Phys. Lett.*, 59 p.2130.
70. Rashkeev, S.N., Di Ventra, M., and Pantelides, S.T. (2001): *Appl. Phys. Lett.*, 78 p.1571.
71. Kondo, M., Yamasaki, S., and Matsuda, A. (2000): *J. Non-Cryst Solids*, 266–269 p.544.
72. Nishi, Y. (1971): *Jpn. J. Appl. Phys.*, 10 p.52.
73. Zhou, J.-H., Baranovski, S.D., Yamasaki, S., Ikuta, K., Kondo, M., Matsuda, A., and Thomas, P. (1998): *Phys. Stat. Sol.* (B), 205 p.147.
74. Street, R.A., and Biegelsen, D.K. (1980): *Solid. State Commun.*, 33, 1159.
75. Takenaka, H., Ogihara, C., and Morigaki, K. (1988): *J. Phys. Soc. J.*, 57 p.3858.
76. Finger, F., Malten, C., Hapke, P., Carius, R., Wagner, H., and Scheib, M. (1994): *Phil. Mag. Lett.*, 70 p.247.
77. Finger, F., Muller, J., and Wagner, H. (1998): *Phil. Mag.* B 7, p.805.
78. Hasegawa, S., Kishi, K., and Kurata, K. (1985): *Phil. Mag.*, B 52.

5

Key Issues for the Efficiency Improvement of Silicon-Based Stacked Solar Cells

Yoshihiro Hamakawa

Aiming for the development of next-generation solar cells having super high efficiency with low cost, a series of R&D studies on a-Si//poly or μc (microcrystalline or nanocrystalline)-Si thin-film stacked solar cells are introduced in this chapter. First, the basic concept of the multibandgap stacked solar cell and its historical background on the selection of material combinations are explained with an optimum design theory. It has been shown from the results of isoenergy mapping of the material combinations that plasma-CVD-produced amorphous silicon (a-Si) for the top cell with polycrystalline silicon (poly-Si) or microcrystalline silicon (μc-Si) would be the best combination of materials in view not only of the value of energy gaps but also the abundance of natural resources with cost.

Secondly, some key issues for efficiency improvement on the a-Si top cell are briefly introduced, and some technical knowledge on these items is demonstrated. Then, a series of experimental approaches along with the issues are presented on the double heterostructure a-Si top cell and optimum design parameters on the a-Si top cell are determined.

Thirdly, current R&D trends in the bottom-cell technologies are classified into two categories with active material thickness, that is, μc-Si having the thickness of 3–10 μm, and thick film of more than 50 μm, and their present status in current technologies are introduced. As the new material for the bottom cell, some new data on the microcrystalline deposition technologies are also overviewed. Then, recent topics on the microcrystalline (μc-)-based bottom cells are reviewed. Finally, a comparison is made on the realistic achievable theoretical efficiency obtained from computer simulation with the best experimental data and remaining problems are discussed.

5.1 Principle of the Stacked Solar Cell

Among the wide variety of solar photovoltaic R&D efforts, improvement of solar-cell performance is of prime importance. In amorphous-silicon solar-cell projects, a wide variety of routes have been used to improve cell efficiency making full use of technologies, such as film-quality improvement, new junction structures with wide gap materials, such as a-SiC, μc-Si, and narrow gap material like a-SiGe, and the use of new electrode materials with the back surface filed (BSF) effect. For example, an a-SiC/a-Si heterojunction solar cell, which first broke through the 8% efficiency barrier in 1980 [1] is a typically successful one, and has now become a routine technology for fabrication of high-efficiency solar cells.

Another possibility for further improvement of amorphous solar-cell efficiency is heterostructure stacked junctions utilizing internal the carrier exchange effects through localized states at the heterojunction interface. This idea was first proposed by an Osaka University group in 1979 [2], and was applied to the high-voltage solar cell called "HOMULAC (horizontally multi-layered photovoltaic cell)" [2]. Figure 5.1 shows the energy band diagram and schematic representation of hole–electron exchange effects at the junction (a) and an equivalent-circuit explanation of the HOMULAC device (c). To obtain efficient photon-energy collection with a multibandgap stacked solar cell, there are some design rules described below.

In the two-terminal stacked solar cell, the i-layer thickness of i-th stacked cell d_i should be designed so that it is also the same photogenerated current I_{Li} as the other i-layer in order to fulfill the current continuity law; that is,

$$I_{Li} \cong q \int \Phi_i(\lambda)\eta_c(\lambda, V_i)\mathrm{d}\lambda = \text{constant} \ , \tag{5.1}$$

where $\Phi_i(\lambda)$ is the incident photon flux density at the wavelength λ into the i-th junction interface, η_c = the photocarrier collection efficiency in the i-th layer. In the simplest treatment, in which both the carrier collection losses and the optical interference effects are neglected, η_c and Φ_i may be approximated by using the absorption coefficient $a_i(\lambda)$ and the thickness of the i-th photovoltaic layer, and by defining the incident photon flux $\Phi_0(\lambda)$ as

$$\eta_c(\lambda) \simeq I - \exp[-\alpha_i(\lambda)d_i] \tag{5.2}$$

and

$$\Phi_i(\lambda) \sim \Phi_0(\lambda) \prod_{j=1}^{i-1} \exp[-\alpha_i(\lambda)d_i] \ . \tag{5.3}$$

These approximations give the maximum limit of I_{Li}. Optimization of the stacked cell is performed by choosing a set of d_i that satisfies the condition $I_{Li} = I_L/m \ (i = 1, \ldots, m)$ [3]. As can be seen in the photon flux penetration pattern shown in Fig. 5.1b, that in order to have an energy gain from the

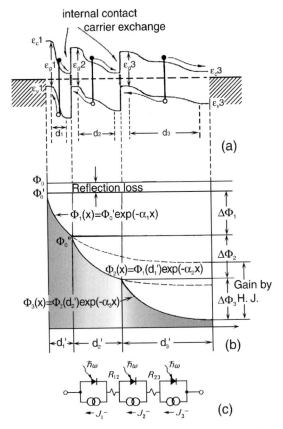

Fig. 5.1. Band diagram explanation of the (a) multi-band gap stacked solar-cell, (b) its photon flux penetration pattern, and (c) equivalent circuit

i-th stacked layer heterostructure junction, the absorption coefficients in both sides of neighboring i-layers should satisfy the relationship

$$\alpha_{i-1}(\lambda) < \alpha_i(\lambda) < \alpha i + 1(\lambda) \; . \tag{5.4}$$

For the selection of material of the i-th layer semiconductor, we need a rule not only for the order of the energy gap $E_{g1} > E_{g2} > E_{g3}$ but also for $\alpha_1(\lambda) < \alpha_2(\lambda) < \alpha_3(\lambda)$ as illustrated in Fig. 5.2.

The I-V characteristics of i-th layer can be written in the form

$$I = I_{Li} - I_{si}\left[\exp\left(\frac{q(V_i + IR_{si})}{n_i kT}\right) - 1\right] - \frac{V_i + IR_{si}}{R_{shi}} \;, \tag{5.5}$$

for $i = 1, \ldots, m$, where V_i, I_{Li}, I_{si}, n_i, R_{si}, and R_{shi} denote the voltage, photocurrent, saturation current, diode ideality factor, series resistance, and shunt resistance related to the i-th layer cell, respectively. It should be noted that the photocurrent I_{Li} depends on Vi because the photocarrier collection

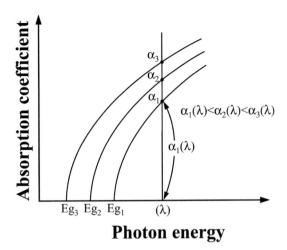

Photon energy

Fig. 5.2. Explanation of the absorption coefficient order rule

across amorphous p-i-n junctions is mainly due to the drift component and thereby is strongly field dependent [4].

Equation (5.5) can be rewritten with the additional simplifications of $n_i = n$ and $R_{shi} = \infty$ for $i = 1, \ldots, m$:

$$V + IR_s = \ln \prod \left(\frac{I_{Li} + I_{si} - 1}{I_{si}} \right) \qquad (5.6)$$

$$\text{with } V = \sum_{i=1}^{m} V_i \text{ and } R_s = \sum_{i=1}^{m} R_{si} \ .$$

It can easily be verified by a simple mathematical treatment of (5.6) that both the short-circuit current Isc and open-circuit voltage V_{oc} of the stacked cell attain their maximum when each photocurrent I_{li} is equal to $1/m$ of the photocurrent I_L defined by the summation of I_{Li}. Under this optional condition (5.6) can be reduced to [5]

$$\bar{V} + \overline{IR_s} \cong \frac{nkT}{q} \ln \left(\frac{I_L + \bar{I}_s - \bar{I}}{\bar{I}_s} \right) \ . \qquad (5.7)$$

with definitions $\bar{V} = V/m$, $\bar{I} = mI$, $\bar{I}_s = m\left(m \prod_{i=1}^{m} I_{st} \right)^{1/m}$ and $\bar{R}_s = R_s/m^2$. Equation (5.7) indicates that the heterojunction stacked cell is equivalent to a single-junction cell characterized by the photocurrent I_L, saturation current I_s, and series resistance R_s, as far as the curve-fill factor and conversion efficiency are concerned. Thus, the maximum obtainable conversion efficiency can be given by using (5.7):

$$\eta_{\max} \cong \frac{\bar{V}_m \bar{I}_L}{P_{in}} \left\{ 1 - \frac{\exp(q\bar{V}_m/nkT) - 1}{\exp(q\bar{V}_{oc}/nkT) - 1} \right\} \ , \qquad (5.8)$$

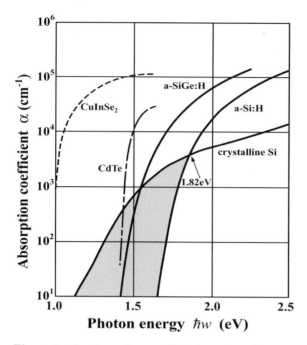

Fig. 5.3. The absorption coefficient spectra of some promising candidate materials for bottom junction stacked with an μ-Si solar cell

where P_{in} is the incident solar power, V_{oc} denotes

$$\bar{V}_{oc} = \frac{nkT}{q} \ln\left(\frac{\bar{I}_L}{\bar{I}_S}\right) \; , \tag{5.9}$$

and V_m is given by the solution of

$$\exp\left(\frac{q\bar{V}_m}{nkT}\right) \cdot \left(I + \frac{q\bar{V}_m}{nkT}\right) = \frac{\bar{I}_m}{\bar{I}_S} + 1 \; . \tag{5.10}$$

On the basis of the procedure outlined above, Tsuda et al. [6] and Fan and Palm [7] have calculated the theoretical limit of the conversion efficiencies for various energy gap material combinations. Figure 5.3 shows the absorption spectra of some well-matched candidate materials stacked with a-Si following both design rules as expressed in (5.1) and (5.4).

Figure 5.4 shows the achievable isoefficiency map of Si-based tandem-type solar cells as a function of the bandgap energy of the top and bottom cell materials reported by Takakura [8]. Calculation has been made on a-Si/poly-Si absorption spectra of a two-terminal stacked solar cell (a) and the four-terminal one (b).

As can be seen in the figure, the crystalline-silicon solar cell is a promising bottom cell for the combination to a-Si top cell. In 1984, the Hamakawa group

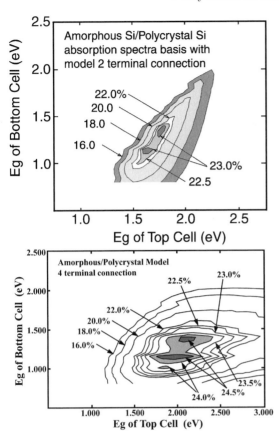

Fig. 5.4. Isoefficiency maps of amorphous/polycrystal model tandem solar cells in (top) the two-terminal and (bottom) four-terminal cells

of Osaka University reported a-Si//poly-Si stacked solar cells by the name of "Honeymoon Cell", which means a marriage of a-Si and c-Si technologies [5].

Significant advantages in this material combination are enumerated as:

i) A good energy gap combination to a-Si having a high theoretical limit of conversion efficiency (23% for a two-terminal cell and 24.5% for a four-terminal cell);

ii) A relatively well-established technology on single and poly-Si thin-film growth with hydrogen passivation;

iii) No Staebler–Wronski effect in the crystalline-Si bottom cell;

iv) Pollution-free material for the material preparations and abundant resources with low cost.

5.2 An Optimum Design of the a-Si Top Cell

There have been tremendous R&D efforts for the efficiency improvement of a-Si solar cells since their invention in 1977. The principal processes to improve the efficiency are enumerated as: (A) an efficient guidance of photon energy into the PV active layer, (B) more effective photon confinement in the PV active layer, (C) more efficient photogenerated carrier confinement, (D) suppression of the recombination losses, (E) reduction of the voltage factor loss, (F) decrease of the series resistance loss. The technological knowledge obtained from the R&D efforts is summarized in Table 5.1.

With full use of the above technological knowledge, an optimum junction structure of the a-Si top cell designed is illustrated in Fig. 5.5. Textured TCO/glass substrate was employed for the utilization of optical-confinement effects. During the fabrication process of the solar cell, as the first step, a p-type μc-SiC electrode layer with a high dark conductivity of 0.1 S cm^{-1} and wide optical bandgap of 2.07 eV were selected to avoid the cross-contamination effects. The substrate temperature during the deposition was fixed at 250°C.

Table 5.1. Key Issues for Efficiency Improvement of a-Si Solar Cells

	Physical process	Technical solution
A	Efficient guidance of optical energy	a-1) Antireflection coating (ARC) a-2) Multi-energy-gap stacked junction
B	Efficiently guided photon confinement	b-1) Textured surface treatment b-2) Use of back-surface-reflection (BSR) effect b-3) Refractive index arrangement
C	Carrier confinement	c-1) Minority carrier mirror effect by heterojunction c-2) Increase of $\mu\tau$-product in the PV active layer
D	Reduction of photogenerated carrier recombination	d-1) Film quality improvement by controlling the deposition condition such as RH, T_s, RF-frequency d-2) Drift-type effect with p-i-n junction d-3) Graded-gap PV active layer (bandgap profiling) d-4) Graded impurity-doping involving back surface field (BSF) effect
E	Reduction of voltage factor losses	e-1) Band profile control of the PV active layer e-2) Insertion of proper buffer layer in the interface of the p-i and i-n junction
F	Reduction of series resistance losses	f-1) Optimum design of electrode pattern f-2) Decrease of transparent conductive oxide (TCO) resistance f-3) Use of superlattice tunneling junction

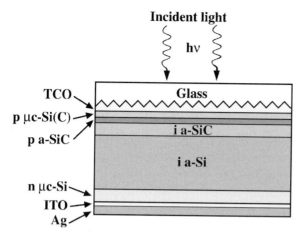

Fig. 5.5. Structure of a-Si double heterojunction

Recently, some intensive investigations are also in progress to explore the best deposition condition of the i-layer [9, 10]. In our case, the thickness of each constituent layer was p a-SiC 10 nm, i a-Si in the range from 100 nm to 200 nm and n μc-Si with a range of 20 to 30 nm with the optical energy gap (determined by Tauc's plot) of 1.89, 1.75, and 1.90 eV, respectively.

With the optimizations of all the constituent layer thicknesses, the best output characteristics of an a-Si single-junction solar cell, was obtained up to now as 12.5% with $V_{oc} = 0.905$ V, $J_{rmsc} = 18.8$ mA cm^{-2}, and FF = 73.6% as shown in Fig. 5.6. Here, it should be noted that there remains room for

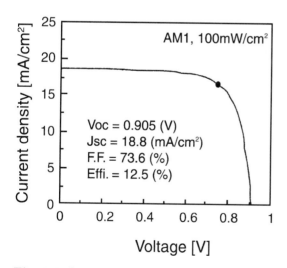

Fig. 5.6. Output characteristic of a-Si solar cell with an n-type c-Si layer inserted between TCO and p-type c-Si

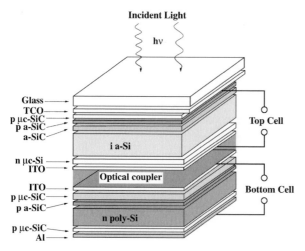

Fig. 5.7. Structure of a-Si//poly-Si four-terminal heterojunction tandem solar cell

improvement in V_{oc} by at least 0.05 V by an extended optimization procedure, particularly concerning the TCO/p-layer/i-layer region.

5.3 Poly-Si and μc-Si Bottom-Cell Technology

To confirm the achievable theoretical efficiency as shown in Fig. 5.4, a series of experimental verifications has been made with a cast-Si bottom cell. Fig. 5.7

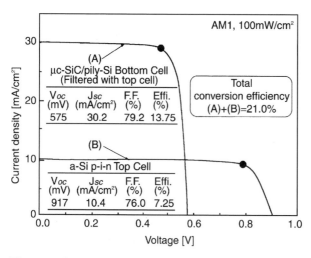

Fig. 5.8. Output characteristics of a-Si(B)//poly-Si(A) four-terminal tandem solar cell

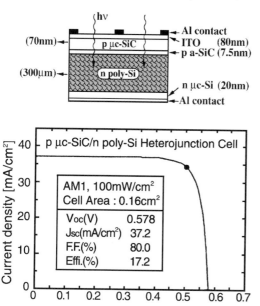

Fig. 5.9. μc-SiC/poly-Si heterojunction solar cell and its output characteristics (presented by Osaka Univ.)

shows the cell structure of a prototype a-Si with poly-Si four-terminal stacked solar cell. It has been shown with this prototype junction structure on four-terminal 2 stacked cell that 21% efficiency is achieved. This is clearly higher than both a-Si and polycrystalline-Si single-junction solar cells, as shown in Fig. 5.8 [11, 12].

During the R&D efforts for the "Honeymoon cell" technology, an interesting byproduct has been produced. That is, the amorphous or microcrystalline (μc-Si)/poly Si heterojunction solar cell as shown in Fig. 5.9 [13]. The main features of this cell are, 1) a very simple processing method, 2) low cost, 3) more than 17% efficiency was easily obtained. Recently, the Sanyo group [14] has reported 21% efficiency with this technology on 1×1 cm^2 area laboratory-phase cell having p-i i-a-Si deposited on an n-c-Si cell structure, called a HIT (heterojunction with intrinsic thin layer) cell. Presently, this technology has grown to mass production with the commercial name of HIT Power 21 having 15.2% module efficiency, which is one of the highest module efficiencies.

On the basis of the "Honeymoon cell" concept, many kinds of material preparation challenges have been started worldwide for the bottom-cell fabrication technologies, for example, ZMR (zone melting recrystallization) [15], lateral graphoepitaxy [3], plasma spray of Si growth (PSSG) [16], solid-phase crystallization (SPC) [17], CAT (hot-wire)-CVD [18] and also microcrystalline-Si thin-film growth by the plasma CVD method [19] and so on. Table 5.2 shows the classification of the bottom-cell technologies with their active layer thickness. Apart from a prototype cell that employed the cast poly-Si, which is used to verify the original idea, the practical bottom-cell

Table 5.2. Bottom-cell technologies in a-Si//poly (μc)-Si stacked solar cells *2T: two-terminal, **4T: four-terminal, ***S.J.: bottom-cell single-junction

Classification	Bottom cell	PV active layer thickness (μm)	R&D efficiency (%)	Institution
I. Prototype	Sliced cast poly-Si	100–250	15.6 for 2T* 21.0 for 4T**	Osaka U. Osaka U.
II. Thick film	Lateral graphoepitaxy	5–10	8.5 S.J.***	Ritsumeikan U. Mitsubishi
	ZMR lateral graphoepitaxy	(3–5)+(50–60)	16.4 S.J.***	Sanyo
	SPC	10		Daido-Hoxan
	PSSG	200–300		
III. Thin film	GD a-Si recrystal	3–5	14.5 for 2T*	Kaneka
	"Micromorph Si"	1–3	9.1 for 2T*	Neuchatel U.
	"VHF-PE-CVD"	3–5	12.0 for 2T*	Jülich U.
	"Polymorphous Si"	3–5		
	PE-CVD produced μc-Si	3	9.28 for 2T*	Ritsumeikan U.
	PE-CVD produced μc-Si	2–5	9.18 S.J.***	/ Osaka U.
	PE-CVD produced μc-Si	3	12.75 S.J.***	Canon
	PE-CVD produced μc-Si	3	9.4 S.J.***	AIST Tsukuba

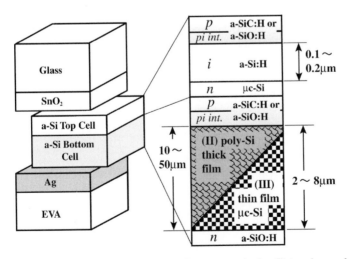

Fig. 5.10. Schematic diagram of a two-stacked a-Si tandem solar cell

thickness can be classified by poly-Si thick film (II) and μc-Si thin film (III), as shown in Fig. 5.10. There is insufficient space to explain the details of these technologies. In the SPC technology, the poly-Si film was grown from a 10-μm thick GD a-Si film deposited on a textured substrate with a high substrate temperature of 500–600°C. The n-type film has an electron Hall mobility of 623 cm^2 V^{-1} s^{-1} with a grain size around 6 μm [17]. With this SPC-produced material, 8.5% efficiency with $Isc = 24.2$ mA cm^{-2} and $V_{\text{oc}} = 0.537$ V are obtained. This result could be evaluated as an excellent performance considering the active layer thickness of only 10 μm poly-Si.

Figure 5.11 shows a schematic illustration of the ZMR method and V-I characteristics of a ZMR poly-Si solar cell. As can been seen in the figure (a), firstly 3–5 μm microcrystalline-silicon (μc-Si) is deposited by LP-CVD on the SiO$_2$-coated silicon substrate, then another SiO$_2$ coating is made on the deposited μc-Si. After that, the ZMR process is carried out by a traveling upper heater to perform a kind of zone melting grain growth. As a final step, a 50–60 μm active layer is formed by the thermal CVD method. A 16.4% efficiency has been obtained so far with a 2×2 cm^2 cell area [20].

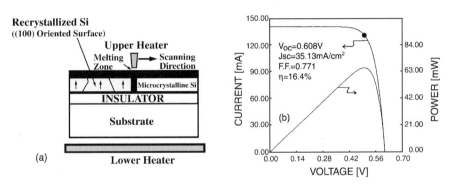

Fig. 5.11. (left) A schematic illustration of ZMR technology and (right) output characteristics of ZMR grown poly-Si solar cell

The third class of technology of the thin-film bottom-cell material might be a kind of microcrystalline Si. As has been reported elsewhere in the recent research on this material, there exist at least two points of significance in this material system. First, as a basic property, μc-Si has a very high optical absorption coefficient in the visible solar radiation region, as has a-Si. In addition, it has a high carrier mobility, at least one order of magnitude higher than a-Si. The second significance is that one can produce the bottom cell by a PE-CVD continuous process together with an a-Si top cell. Along with this concept, a two-dimensional computer simulation study has also been made on the cell structure, as shown in Fig. 5.12 (right). With suitable material parameters and a 10-μm thick active layer, 18.4% efficiency has been obtained, as shown in Fig. 5.12 (left) [21].

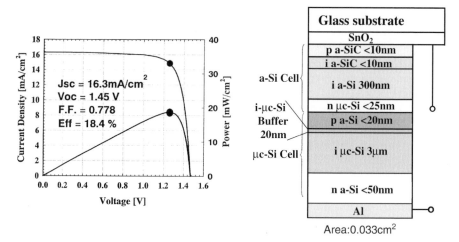

Fig. 5.12. (left) Realistically achievable output characteristics of μc-Si basis bottom solar cell by 2D computer simulation and (right) its cell structure of a-Si//μc-Si stacked solar cell

On the other hand, in the experimental trials, many types of microcrystalline Si growth experiments have been in progress recently, termed "micromorph" [18] and "polymorphous" [19], etc. In these experimental approaches, the size and shape of the μc-Si are very sensitive to the deposition parameters, such as hydrogen dilution ratio R, substrate temperature T_s, injected VHF power P [22]. There are still many discussions on these items related to the film quality and the obtained PV performance [23]. An important key tech-

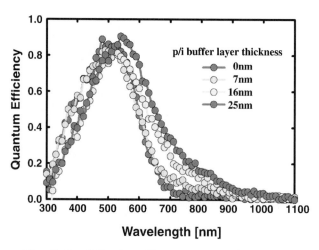

Fig. 5.13. Collection-efficiency curves of bottom cell in the structure of p-a-Si/I-μc-Si as a parameter of the buffer-layer thickness

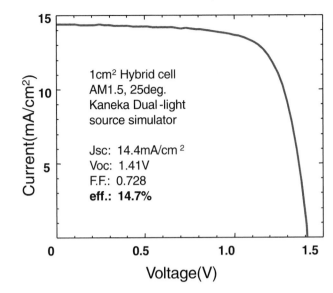

Fig. 5.14. V-I characteristic of a-Si//μc-Si stacked solar cell

nology exists around a transition region from a-Si to μc-Si growth condition. Recent data on the high-quality μc-Si for solar cells reported by the USSC group supported this idea. Recently, the Ritsumeikan group [21] developed a new technique to improve the i-μc-Si layer crystallinity introducing an i-μc-Si buffer layer deposited by PV-CVD under low-pressure conditions. The junction of the structure of a-Si//μc-Si stacked solar-cell shown in Fig. 5.12b is well examined with the collection efficiency curves of the bottom cell in the structure shown in Fig. 5.13 as a parameter of the buffer layer thickness. As can be seen in this figure, the p/i buffer layer thickness is an important key factor in the formation of the μc-Si active-layer quality. Among these approaches, the KANEKA group has reported recently the world best efficiency of 14.5%, as shown in Fig. 5.14 [24] and [25].

References

1. Hamakawa, Y., and Tawada, Y. (1981): *Int. J. Solar Energy*, 1, No.1 pp.125–144, and US Patent 4, 385, 200, Aug. 30 and 5 others.
2. Hamakawa, Y., Okamoto, H., and Nitta, Y. (1979): *Appl. Phys. Lett.*, 35 p.187 and US Patent 4, 271, 328, Jan. 2, 1981.
3. Hamakawa, Y., Matsumoto, Y., Xu, C. Y., Okuyama, M., Takakura, H., and Okamoto, H. (1986): *Mater. Res. Soc. Symp. Proc.*, 70 p.481.
4. Okamoto, H. (1981, 1983): p.94.
5. Hamakawa, Y., Okamoto, H., and Y. Nitta (1984): Proc. 14th IEEE PVSC p.1074.
6. Tsuda, S., Kuwano Y., et al. (1982): *Jpn. J. Appl. Phys.*, 21, Suppl. 21-2 p.251.

7. Fan, J.C.C., and Palm, B.J. (1983): *Solar Cells*, 10 p.81.
8. Takakura, H. (1992): *Jpn. J. Appl. Phys.*, 31 pp.2394–2399.
9. Koch, C., Ito, M., Svrcek, V., Schubert, M.B., and Werner, J.H. (2000): MRS Spring Mtg. San Fransisco, A15.6.
10. Yang, J., Lord, K., and Guha, S. (2000): MRS Spring Mtg. San Fransisco, A15.4.
11. Ma, W., Horiuchi, T., Okamoto, H., and Hamakawa, Y. (New Delhi, 1992): Tech. Digest of PVSEC-6 p.463.
12. Ma, W., Horiuchi, T., Lim, C.C., Yoshimi, M., Okamoto, H., and Hamakawa, Y. (1993): Proc. 23rd IEEE PVSC p.338.
13. Yoshimi, M., Ma, W., Horiuchi, T., Okamoto, H., and Hamakawa, Y. (1992): MRS Spring Mtg., Symp. A p.845.
14. Tanaka, M. Taguchi, M., Matsuyama, T., Sawada, T., Tsuda, A., Nakano, S., Hamafusa, H., and Kuwano, Y. (1992): *Jpn. J. Appl. Phys.* 31, p.3518.
15. Takami, A., Arimoto, S, Morikawa, H., Hamamoto, S., Ishihara, T., Kumabe, H., and Murotan, T. (Amsterdam, 1994): Proc. 12th EC-PVSEC p.59.
16. Kumagai, K. MITI (June, 1994): Report of 31st Solar Energy Tech. Promotion Committee, New Sunshine Project HQ, p.83.
17. Matsuyama, T., Baba, T., Takahama, T., Wakisaka, K., and Tsuda, S. (1994): *Optoelectron. Devices, and Techn.*, 9 p.391.
18. Schropp, R.E.I., et al. (Sapporo, 1999): 11th PVSEC p.929.
19. Meier, J., et al. (Sapporo, 1999): 11th PVSEC p.221.
20. Morikawa, H., Nishimoto, Y., Naomoto, H., Kawama, Y., Takami, A., Arimoto, S., Ishihara, K., and Namba, K. (1998): *Solar Energy Mater. Solar Cells* 53 pp.23–28.
21. Hamakawa, Y., and Takakura, H. (2000): Proc. 28th. IEEE, Alaska pp.766–771.
22. Shah, A., Platz, R., Fischer, D., Hof, C., Ziegler, Y., and Goerlitzer, M. (Tokyo, 1999): Tech. Digest of Review Meeting for Research with Overseas Country Research Inst., pp.9–16.
23. Roca, P., and Cabarrocas, I. (1999): Abstr. 18th Int. Conf. Amorphous and Microcrystalline Semiconductors, Snowbird, p.210.
24. Yamamoto, K., Yoshimi, M., and Tawada, Y.: Tech. Digest of WCPEC-III, Osaka, Japan (2003) S20-B9-03.
25. Fukuda, J., and Yamamoto, K. (2002): 1st PVTEC Technical Forum pp.1–5.
26. Nasuno, Y., Kondo, M., and Matsuda, A. (2001): *Jpn. J. Appl. Phys.*, 40 p.303.
27. Ogawa, K., Saito, K., Sano, M., Sakai, A., and Matsuda, K. (2001): Tech. Digest, PVSEC-12, Jeju, Korea, pp.343–346.

6

Development of Amorphous-Silicon Single-Junction Solar Cells and Their Application Systems

Katsuhiko Nomoto and Takashi Tomita

Worldwide consciousness of the global environmental issues creates rapidly growing demands for solar photovoltaic systems. Sharp has increased the production capacity of solar cells and modules by about 90-fold in the last eight years. To date, the products are mainly bulk-type crystalline-Si solar cells, but to meet the rapid growth of the market it is essentially important to establish new technologies based on the concept of reduced use of material silicon.

For this purpose, we have been developing the technologies of large-scale, thin-film-type amorphous-Si (a-Si:H) solar cells and up to now stabilized cell conversion efficiencies of 7.5% with a monolithic a-Si:H single-junction structure, and size of 730 × 980 mm, are obtained.

In this chapter the key manufacturing technologies and the device design for large-scale, high-efficiency a-Si:H single-junction solar cells are described. Also, our unique application systems of the a-Si:H solar modules towards further expansion of the PV market are introduced and illustrated.

6.1 Introduction

Intensive research and development of thin-film solar cells is being made all over the world, in addition to the current bulk-type, crystalline-Si (c-Si) solar cells, to target the rapidly growing PV market [1]. Although amorphous Si (a-Si:H)-type solar cells have been considered one of the most hopeful thin-film solar cells [2], the market requires even more cost competitiveness and unique applications, as well as higher cell performance.

From this standpoint we have focused our efforts on the development of the manufacturing technologies of the monolithic, large-scale, single-junction a-Si:H solar cells and of unique modules and system applications that clearly differentiates c-Si modules.

This chapter elucidates two important technologies of the large-scale a-Si:H solar cells that we have developed in terms of the manufacturing process and device structure. One is the growth technology of device-quality a-Si:H films depositing on large-scale substrates, with mass-production throughput, named the short-pulse VHF plasma CVD (S·VHF p-CVD) method [3]. The other is a p-type window-layer material of wide bandgap a-Si:H [4], which is critically important for the device performance. Based on these technologies a stabilized conversion efficiency of 7.5% with a monolithic a-Si:H single-junction structure, substrate size of 730 × 980 mm, has been obtained so far.

On the other hand, for further expansion of the PV market, other than the market of the residential rooftop applications, the main market today, it is very important to develop technologies applicable to various spaces of public buildings, the walls of industrial factories, or architectural structures. In such applications, a-Si:H PV modules are suitable because of the increase in available space over the residential rooftop and easy scale-up of a-Si:H solar cells that also lead to cost reduction of the PV modules. In this chapter such an example of our a-Si:H PV modules for the industrial use, which are directly integrated onto the construction material of the autoclaved light-weight concrete, abbreviated ALC by Sumitomo Metal Mining Co., Ltd., is described. Another good example of the application is the architectural, design-oriented use of a-Si:H thin-film modules. A typical example is a see-through a-Si:H module that features the function of light-through as well as light-to-electricity conversion. Representative application systems of these unique a-Si:H modules are illustrated.

6.2 Key Technologies and Approaches Towards Large-Scale, High-Efficiency, a-Si:H Single-Junction Solar Cells

6.2.1 Basic Cell Structure and Process

Our basic structure of a single-junction a-Si:H solar cell is shown in Fig. 6.1 together with the integrated cell structure. Figure 6.2 shows the schematic manufacturing process flow.

We use a glass substrate covered with textured tin oxide. Onto the substrate, pin single-junction layers of a-Si:H are deposited by a glow-discharge S·VHF-plasma CVD system with cathode electrodes of about 1 m² size. These a-Si:H pin layers are scribed and divided into the unit cells by laser processing, following the formation of the rear contact, which consists of a film of a transparent conductive oxide and a high refractive metal, via sputtering. In this process, each unit cell is automatically series-connected, producing a large-scale integrated a-Si:H solar cell. After this cell process, a-Si:H modules are fabricated by a wire bonding, a back-sheet lamination process, and final characterization of the I-V performance of every module, as is shown in Fig. 6.2.

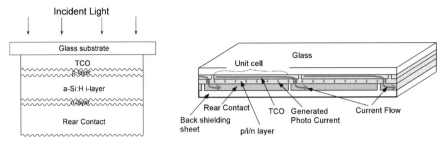

Fig. 6.1. (left) Structure of a-Si:H single-junction solar cell; (right) Structure of a-Si:H single-junction, monolithic, integrated cell

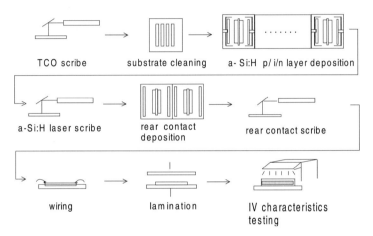

Fig. 6.2. Process flow of a-Si:H single-junction solar cell and module

Among the processes mentioned above, a key issue is to control the processing and properties of the intrinsic a-Si:H layer from the viewpoint of massproduction. As to the device structure, the light-incident p-layer is one of the most important factors to obtain high conversion efficiency of a-Si:H single-junction solar cells [6]. Some detail description concerning these technologies is given in the next section.

6.2.2 Key Manufacturing Technology and Device Design

In order to control the processing and film properties of the intrinsic a-Si:H during mass production, we have developed a novel short-pulsed VHF plasma-enhanced chemical vapor deposition method (S·VHF p-CVD), based on the study of characterization of film properties and plasma diagnostic analysis [3].

It has been reported that low-frequency modulation of the RF discharge for chemical vapor deposition of a-Si:H films with periods of the plasma-on and plasma-off times of a few hundred microseconds (200–500 μs) successfully suppressed powder particles in the discharge space [5]. Figure 6.3 shows a

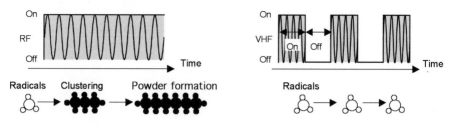

Fig. 6.3. Schematic illustration of the effect of the pulsed plasma

schematic explanation of the effect of the pulsed plasma. Continuous plasma discharge of a SiH_4/H_2 gas mixture under conditions of a high density of SiH_x radicals produces high deposition rate of a-Si:H, which produces powder particles of polysilane. These particles not only deteriorate film electrical properties of a-Si:H but also lead to frequent chamber cleaning that markedly reduces production efficiency. Pulsed plasma of the order of a few hundred microseconds, on the other hand, dramatically reduces such particle production.

Furthermore, plasma kinetics considerations have led us to find that in a much shorter time scale the structural and electrical properties of a-Si:H films can be improved. In this case we use very high frequency (VHF) (more than 27.12 MHz) plasma excitation because much higher radical densities have to be produced in a very short plasma-on time (1–$10~\mu s$) compared to the case of cw discharge and the required high electron densities for the dissociation of source gases are expected to be obtained by VHF excitation.

Figure 6.4 shows photoconductivity (σ_{ph}) to dark conductivity (σ_d) ratios versus the pulsed plasma on-time, together with the variation of the content of the dihydride mode Si-H_2 in the films. This figure clearly indicates that the σ_{ph}/σ_d ratio improves in the region of the plasma-on time less than $10~\mu s$, reflecting a structural change of the file, i.e., the decrease of the dihydride mode that is thought to cause deterioration of film properties. Using this technology, good film quality of the i-layer and suppression of the powder particles in the discharge space are simultaneously realized under the conditions of mass production.

To obtain high conversion efficiency a-Si:H solar cell and thereby high-power modules, it is very important to use a wide optical gap and highly conductive material for the light-incident p-type window layer with uniformity over large areas. We have proposed the "hydrogen plasma doping" method [4] to make such p-type a-Si:H films without alloying a-Si:H with carbon or oxygen (a-SiC:H, a-SiO:H) [6, 7]. We named it WG-p (wide gap p-type) film and the process flow is shown in Fig. 6.5. From the comparison of the spectral collection efficiency of single-junction pin a-Si:H solar cells between the cell with the WG-p layer and the one with a conventional a-SiC:H p-layer in

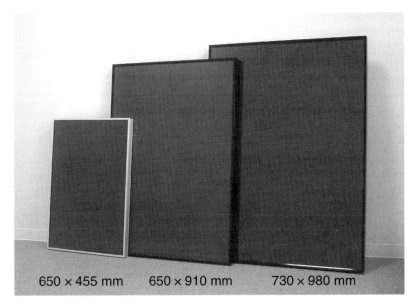

Fig. 6.8. Typical a-Si:H single-junction modules

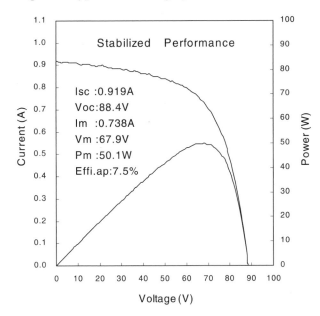

Fig. 6.9. I-V parameters and curve of 730×980 mm module

The dependences of the output power on irradiation light intensity and module temperature are shown in Fig. 6.10 and Fig. 6.11 respectively. The power of our single-junction a-Si:H module is linearly dependent on the irra-

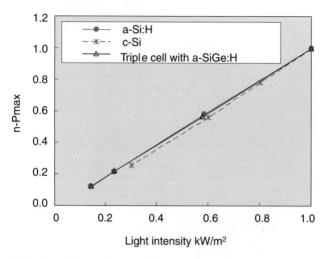

Fig. 6.10. Dependence of the output power on irradiation light intensity

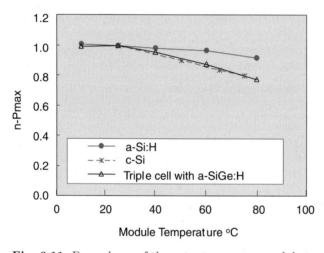

Fig. 6.11. Dependence of the output power on module temperature

diation light intensity, showing almost the same behavior as for other modules of c-Si type and amorphous multijunction solar cells using a-SiGe:H. The dependence on the module temperature, however, is very different among these modules. Figure 6.11 clearly indicates that the temperature dependence of a-Si:H single-junction solar cells is much smaller than the others and the reduction of the power is only about 5% around at 60–80°C, at which temperatures solar modules operate outdoors. This power reduction is about one-third compared to c-Si type and amorphous multijunction solar cells using a-SiGe:H. This is due to the inherent difference of the bandgap of the semiconductors that convert light to electricity.

As it is well known, amorphous-silicon-related materials, a-Si:H, a-SiGe:H, a-SiC:H, etc., initially change their electrical properties during light irradiation, the so-called Staebler–Wronski effect [8]. This effect also causes initial light degradation of the a-Si:H solar cells. In Fig. 6.12 the power change of our a-SiH single-junction submodule is plotted against light irradiation times. The irradiation light source is the solar simulator and the condition is continuous irradiation, 1 sun: AM.1.5, 1 kW m^{-2}, 50°C (accelerating light irradiation test). The power stabilizes after a few hundred hours irradiation and sustains around 76% of the initial value. Figure 6.13 is the outdoor reliability test data

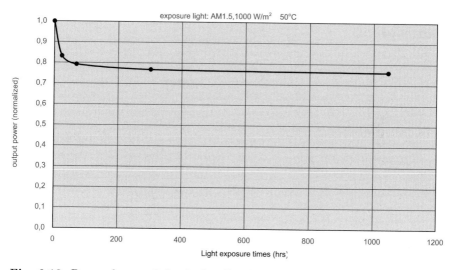

Fig. 6.12. Power change of the single-cell submodule under continuous light irradiation of AM1.5, 1 kW m^{-2}

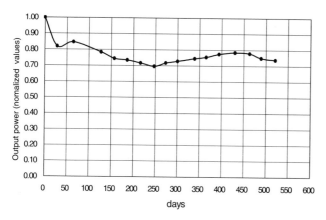

Fig. 6.13. Power change of a-Si:H single-junction solar module against outdoor exposure

and the normalized output powers of the modules are plotted against outdoor exposure days. The power of the module initially degrades and then it varies with the season due to the annealing effect in the high summer temperatures. The module power in this seasonal changing lies within 70% to 80% of the initial value and the average value in this mode corresponds to the value of the stabilized power of the accelerating test in Fig. 6.12. We define, therefore, the module power as the stabilized value according to the accelerating test of the light irradiation of 1 sun and continuous 1000 h irradiation.

6.3 Applications of Large-Scale a-Si:H Solar Modules and Systems

NEF (New Energy Foundation in Japan) leadership has led to the nation-wide spread of the photovoltaic power-generation systems in Japan, which are mainly the residential rooftop applications. For further expansion of the PV market, however, it is very important to develop various application systems that utilize the walls of industrial factories, public buildings, or architectural structures. In such applications, a-Si:H PV modules are suitable because of the increased available space compared to the residential rooftop and the easy scale-up of a-Si:H solar cells. From this point of view two unique modules are described and three application systems are illustrated here.

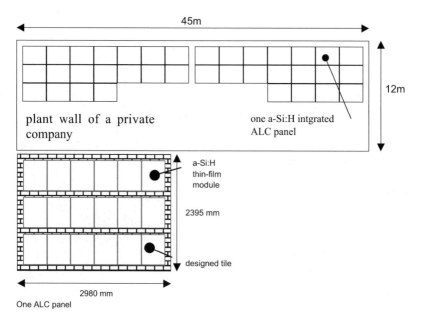

Fig. 6.14. Illustration of plant wall installed a-Si:H integrated ALC panel and one ALC panel integrating a-Si:H solar modules

6.3.1 Application to the Construction Material, ALC Panel Integrating a-Si:H PV Modules

Autoclaved lightweight concrete (abbreviated ALC) is a widely used, inexpensive construction material and there is an estimate of an installation potential of 100 M/y PV modules on the ALC panels. Sumitomo Metal Mining Co., Ltd. is a major manufacturer of ALC material in Japan and the company has developed a-Si:H PV integrated ALC modules as a joint research program with NEDO using our a-Si:H modules. a-Si:H PV modules were chosen by

Table 6.1. Specifications of the PV-integrated ALC panel

Solar cell	: a-Si:H single junction
module size	: 650 × 455 mm
ALC panel	: one string with 6 modules in series, 3 strings in parallel in each ALC panel
array configuraton	: 36 ALC panels
power capacity, voltage	: 10 kW, 240 V

Fig. 6.15. Plant wall of Sumitomo Metal Mining Co. Ltd installing a-Si:H solar module integrated ACL panels with 10 kW capacity

the company considering the temperature increase due to integration of PV modules directly on the ALC panel.

Figure 6.14 shows the structure of the a-Si:H PV integrated ALC module. One ALC panel consist of six sheets of series-connected a-Si:H modules that form one PV string, an a-Si:H module with a size of 650 × 455 mm, and three parallel-connected PV strings. Thus, one ALC panel, with a size of 2345 × 2980 mm, has eighteen sheets of the a-Si:H module. The surface of the ALC panel is engraved in advance for the connector boxes and the cables of the PV modules. Then the a-Si:H modules are easily fixed on the ALC panel with silicone sealing compound. Table 6.1 summarizes these specifications of the PV-integrated ALC panels. By using the structure and the process mentioned above, Sumitome Metal Mining Co., Ltd. has established the technology of in-house fabrication of ALC panel integrating a-Si:H PV modules. An array consisting of thirty-six integrated ALC panels with 10 kW capacity was installed on the wall of their plant in Japan. The picture is shown in Fig. 6.15.

6.3.2 See-Through-Type a-Si:H Solar Modules

Nowadays, architects pay much attention to photovoltaic power systems because of the client requirements for the PV installation to their building or construction bodies, namely highway fences, corridor rooftops and so on. In such cases they are interested not only in the output power of the system but also the design itself of the PV systems harmonizing with the architecture. Here, we describe our technologies of architectural, design-oriented see-through-type a-Si:H solar module and its application systems.

Figure 6.16 shows the structure of our see-through-type module. The same laser scriber as mentioned in Fig. 6.2 is used for see-through patterning, thereby any see-through patterns can be easily realized using corresponding mask patterns if necessary. For instance, according to the customer's special requirements, the modules with see-through patterns of the users' logos have been shipped.

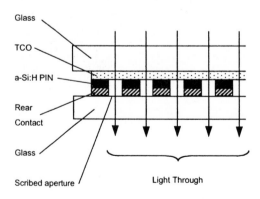

Fig. 6.16. Structure of laser-scribed see-through a-Si:H module

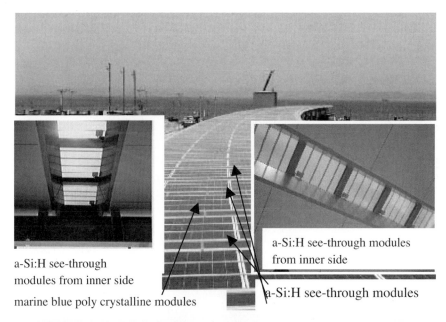

a-Si:H see-through
modules from inner side

marine blue poly crystalline modules

a-Si:H see-through modules
from inner side

a-Si:H see-through modules

Over all scene of the corridor
with a-Si:H see-through modules

Fig. 6.17. a-Si:H see-through module application system with 8.4 kW capacity at the port of Takamatsu city in Shikoku island southwest Japan

Canopy installed a-Si:H
see-through modules

Fig. 6.18. a-Si:H see-through modules with 1.7 kW capacity applied to the canopy
of the new clean room facility of FUJIPREAM, LTD. in Hyogo Prefecture in Japan

In normal cases one can make the see-through lines vertical or parallel to the scribed lines of the rear contact, taking the series resistance loss into account. One can obtain various aperture ratios by changing the number or width of the see-through scribe lines. The rear side of the module is laminated with a glass sheet. Up to now, see-through modules with 10–30% aperture ratios have been commercialized.

The pictures of the actual application system of our see-through module for the corridor roof-top at the port of Takamatsu city in Shikoku island, southwest Japan, are shown in Fig. 6.17. The architect designed this PV system with the purpose of creating the image of marine blue with light sparkling and adopted the polycrystalline Si modules for a marine blue image, 95.1 kW capacity, and a-Si:H see-through modules for entering natural light into the corridor, 8.4 kW capacity. One see-through unit has nine series-connected modules, each module of size 650×455 mm, 10% aperture ratio, output power of 12 W. Each see-through unit is arrayed next to to the marine blue polycrystalline modules, as shown in Fig. 6.17, and the whole system has seventy-eight see-through units. Thus, seven hundred two see-through modules are used in total.

Figure 6.18 shows another example of this type of module applied to the entrance canopy of the new clean room facility of FUJIPREAM., LTD in Himeji city of Hyogo prefecture in Japan. In this case, c-Si-type light-through modules and a-Si:H see-through modules are combined to produce 21 kW of electricity as well as allowing natural sun-light to enter. As to the see-through structure, four a-Si:H see-through modules, each module of size 650×455 mm, 10% aperture ratio, are sandwiched with two tempered glass sheets on the front and rear sides forming one see-through unit of the size of 1400×1000 mm with the output power of 43 W. Forty of these see-through units, with 1.7 kW capacity, are installed on the canopy.

6.4 Conclusion

In summary, we have described our approaches towards largescale, high-efficiency a-Si:H single-junction solar cells to meet the market requirements of more cost competitiveness of PV modules and unique system applications that enlarge PV installation space beyond the residential rooftop.

The key manufacturing technology and the device design of large-scale a-Si:H single-junction solar cells were elucidated, i.e., the short pulse VHF plasma-CVD method and the application of newly developed material of wide-bandgap p-type a-Si:H. Based on these manufacturing technologies, stabilized cell conversion efficiencies of 7.5% with monolithic a-Si:H single-junction structure, with a size of 730×980 mm, are obtained so far.

For further expansion of the PV market, unique a-Si:H modules and application systems that differentiates c-Si type modules were developed; a-Si:H integrated ALC panels were installed on the plant wall of a private company and laser-scribed see-through a-Si:H modules were applied to the design-oriented architectures. These outlines were introduced and illustrated.

Acknowledgment

The authors express grateful acknowledgment to the kind acceptance of the descriptions of their systems of Sumitomo Metal Mining Co., Ltd., Takamatsu city officials, and FUJIPREAM., LTD. Also, the authors would like to express appreciation to the efforts and cooperation of all our project members for the individual technology establishment.

References

1. Hamakawa, Y. (ed.) (2000): Photovoltaic Power Generation. (CMC Books, Japan).
2. Carlson, D.E., Wronski, C.R. (1976): *Appl. Phys. Lett.* 28: pp.671–673.
3. Nomoto, K., Sakai, O., Takagi, S., Sannnomiya, H., Yamamoto, Y., Tomita, T. (1997): Short-Pulse VHF Plasma-Enhanced CVD of High-Deposition-Rate a-Si:H Films. 14th European Photovoltaic Solar Energy Conference and Exhibition, pp.1226–1229.
4. Kishimoto, K., Nakano, T., Itoh, Y., Mashima, S., Sannomiya, H., Nomoto, K. (1999): High Quality p-type Wide Gap a-Si:H Films by Hydrogen Plasma Doping Method. 11th International Photovoltaic Science and Engineering, pp.187–188.
5. Watanabe, Y., Shiratani, M., and Makino, H. (1990): *Appl. Phys. Lett.* 57: pp.1616–1618.
6. Tawada, T., Kondo, M., Okamoto, H., and Hamakawa, Y. (1982): *Solar Energy Mater.* 6: pp.299–315.
7. Fujikake, S., Ohta, H., Asano, A., Ichikawa, Y., and Sakai, H. (1992): *Proc. Mater. Res. Soc. Symp.* 258: pp.875–880.
8. Staebler, D.L., and Wronski, C.R. (1980): *J. Appl. Phys.* 51: pp.3262–3268.

Production of a-Si:H/a-SiGe:H/a-SiGe:H Stacked Solar-Cell Modules and Their Applications

Keishi Saito, Tomonori Nishimoto, Ryo Hayashi,
Kimitoshi Fukae, and Kyosuke Ogawa

In 1996, Canon started mass producing the a-Si:H/a-SiGe:H/a-SiGe:H stacked solar-cell module with over 8% stabilized efficiency, with a unique approach where the microwave-plasma chemical vapor deposition (MW-PCVD) method was implemented in the high-throughput roll-to-roll process. This chapter consists of two distinct parts. The first part describes the R&D effort made for high-efficiency, small-area (laboratory-size) cells. The second part is given to depict the outlines of mass-production method of the solar cell and the products for rooftop applications.

7.1 R&D Work with Small-Area Cells

As the world is becoming more and more seriously concerned about the global environmental issues, photovoltaic (PV) power generation is expected to play a much more active role in the near future [1]. Amorphous-silicon (a-Si)-based thin-film solar cells are a promising candidate for the low-cost power-generation source in the near future, because they are easily formed on a large-area substrate and they require a far smaller amount of silicon source than the crystalline-silicon (c-Si) counterpart. But the current situation is that a-Si-based solar cells are still far from being widely used because their production cost is not yet low enough. The following are thought to be the technological key issues for further cost reduction.

1. Higher conversion efficiency.
2. Higher deposition rate.
3. More efficient source-gas utilization.

To improve the efficiency of the solar cell, the multijunction configuration is useful. An example is the a-Si:H/a-SiGe:H/a-SiGe:H triple-junction stacked solar cell, which can effectively absorb incident light over a wide wavelength range. In the triple-junction configuration, the thickness of the photoactive i layer of each component cell is usually smaller than that of the single-junction cell. This is advantageous for suppressing the light-induced degradation of the cell.

To achieve a higher deposition rate and more efficient source-gas utilization, a high-density and low electron-temperature plasma is required. The 2.45-GHz microwave generator is the highest-frequency plasma source that is commercially available at a realistic cost. Therefore, it has a high potential in achieving the above-mentioned requirements at low cost. The microwave-plasma CVD (MW-PCVD) method was employed for semiconductor layer deposition and an extensive effort has been made to obtain high-quality films at very high deposition rate and at high gas-utilization ratio [2, 3].

The following section describes some aspects of the fundamental R&D work on the small-area (laboratory-size, 0.25 cm^2) cell.

7.2 Low-Pressure Microwave PCVD Method

In the Canon Ecology R&D Center, fundamental R&D work on the a-Si:H/a-SiGe:H/a-SiGe:H stacked solar cell has been in progress for 1. higher conversion efficiency 2. higher deposition rate 3. more efficient source-gas utilization as

1. High-quality a-Si-based semiconductor films are usually obtained at very low deposition rates (0.1–0.5 nm s^{-1}).
2. More pronounced light-induced degradation is observed in a-SiGe:H films than in a-Si:H films. The poor stability of a-SiGe:H films gets even worse when the deposition rate is increased.
3. The GeH$_4$ gas source is very expensive.

Generally speaking, it is not difficult to obtain a high deposition rate if one uses plenty of source gas with high input power. However, this condition often leads to higher silane or powder formation in the plasma, which affects the resulting film quality and device performance. Use of low pressure is one way to reduce gas-phase reactions and the resulting higher silane formation. The low-pressure microwave plasma CVD (LP-MWPCVD) method, which utilizes the 2.45 GHz microwave generator as a plasma source, can make possible continuous discharge at low working pressures on the order of several mTorr and can generate a high density, low electron-temperature plasma. Consequently, with the microwave method, very high deposition rates (more than one order of magnitude larger than in the conventional RF-CVD method) and a high gas utilization ratio (several times higher than in RF-CVD) are easily achieved. The dependence of the deposition rate on silane gas flow rate and microwave power was examined. The results are shown in Fig. 7.1 and Fig. 7.2, respectively.

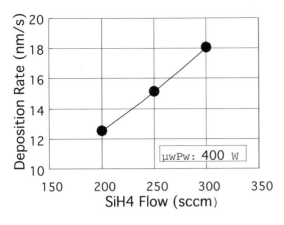

Fig. 7.1. Dependence of the deposition rate on SiH$_4$ flow

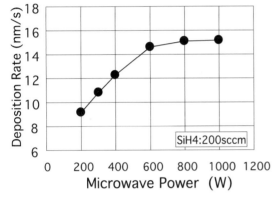

Fig. 7.2. Dependence of the deposition rate on MW power

Estimations from both figures reveal that the silane gas dissociation ratio is close to 100%, which is an amazingly high value considering the fact that 15% is a typical dissociation ratio in RF-CVD. The low-pressure microwave CVD method is expected to be very useful for the deposition of a-SiGe:H layers because the germane gas source is very expensive and it must be used as efficiently as possible. Prior to the mass production of the triple-junction cell, an extensive effort was made in developing small-area cells with the laboratory-size, batch-type apparatus.

Figure 7.3 shows the structure of the small-area cells made with the batch-type process. As shown in Fig. 7.3a, the single-junction cell consists of Grid/ITO/p (μc-Si:H)/i (a-SiGe:H)/n (a-Si:H)/texture ZnO/Ag/stainless steel substrate. The triple-junction cell in Fig. 7.3b consists of Grid/ITO/a-Si:H cell/a-SiGe:H cell/a-SiGe:H cell/texture ZnO/Ag/stainless steel substrate.

Figure 7.4 shows the schematic diagram of the batch-type CVD apparatus for small-area cells. The microwave CVD method was employed solely for a-SiGe:H deposition; the other semiconductor layers of the solar cell were deposited by RF. The substrate is mounted on the load chamber and conveyed to the RF chamber. After the n-layer deposition by RF, the substrate is trans-

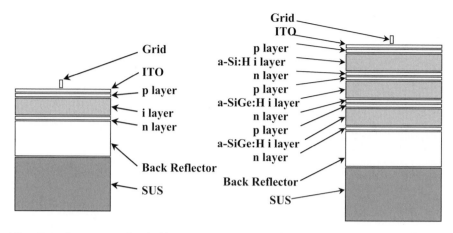

Fig. 7.3. Structures of a (left) single-junction cell, (right) triple-junction cell

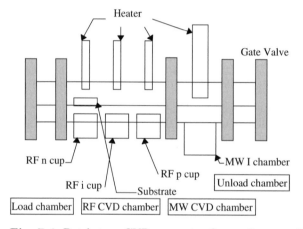

Fig. 7.4. Batch-type CVD apparatus for small-area cell preparation

ferred to the microwave chamber. For the a-SiGe:H photoactive i layer deposition by microwave, a mixture of SiH_4, GeH_4, and H_2 gases was used. The substrate temperature during a-SiGe:H layer deposition was between $200-400°C$. After the a-SiGe:H layer formation is completed, the substrate is moved back to the RF chamber and the p layer is formed. The completed cell was removed from the apparatus through the unload chamber. Other deposition conditions are summarized in Table 7.1. Evaluation of the solar-cell characteristics was conducted under an AM1.5, $100\,mW\,cm^{-2}$ light at a cell temperature of $25°C$.

Table 7.1. Deposition recipes for MW-PCVD and RF-CVD

Receipe for the low pressure MW PCVD
  MW power: $100 \sim 1000$ W
  Pressure: $0.1 \sim 30$ mTorr
  $T_S : 200 \sim 400°$C
  Deposition rate: > 4.0 nm/s
  Source gases: SiH_4, GeH_4, and H_2
Receipe for the RF PCVD
  RF power: $0.1 \sim 10$ Torr
  Pressure: $0.1 \sim 30$ mTorr
  $T_S : 100 \sim 400°$C
  Deposition rate: > 0.1 nm/s
  Source gases: SiH_4, H_2, PH_3, and BF_3

7.3 Graded-Bandgap Profiling in a-SiGe:H

The narrow-bandgap a-SiGe:H material has a tendency to a sharp increase of gap states with the increase of Ge content, which results in a lower photo conductivity and other deteriorated film properties. On the other hand, a-SiGe:H has the advantage that its bandgap can be easily and continuously controlled by changing the mixture ratio of Si/Ge content. This is due to the fact that both Si and Ge belong to the group IV of the periodic table. Therefore, one can utilize the so-called graded-bandgap profiling by changing the Si/Ge content ratio, and can control a-SiGe:H solar-cell properties.

Figure 7.5 shows a comparison of the performance of a-SiGe:H cells with and without graded-bandgap profiling. As can be seen in the figure, the graded cell has larger bandgaps near the n/i and p/i interfaces so as to avoid deteriorated interface properties from bandgap discontinuities. The smallest-bandgap region was intentionally set near the p-layer side to improve the collection efficiency of the photo-generated holes, which limits the device performance. By employing the graded-bandgap profiling, the performance of the a-SiGe:H cell was successfully improved. Both cells shown in Fig. 7.5 were made by the LP-MWPCVD method at a very high rate of about 10 nm s^{-1}.

On the basis of these results, the graded-bandgap profiling technologies are applied to the bottom and middle cells of the triple cell shown in Fig. 7.3b. As has been reported in the literature [2], an optimized triple cell with high-rate (~ 10 nm s^{-1}) a-SiGe:H component cells exhibited a high initial efficiency of 12.3% ($V_{oc} = 2.30$ V, $J_{sc} = 7.38$ mA cm^{-2}, FF $= 0.725$), which was further improved to 12.7% ($V_{oc} = 2.35$ V, $J_{sc} = 7.38$ mA cm^{-2}, FF $= 0.73$) later. The highest initial efficiency obtained with a high rate (> 4 nm s^{-1}) a-SiGe:H i-layers is 13.0% ($V_{oc} = 2.17$ V, $J_{sc} = 8.04$ mA cm^{-2}, FF $= 0.744$) [4] as shown in Fig. 7.6.

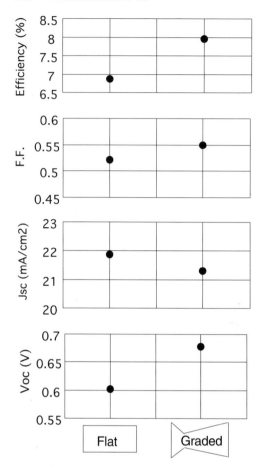

Fig. 7.5. Performance comparison of flat-bandgap and graded-bandgap cell
(SiH$_4$: 100 sccm, GeH$_4$: 50 sccm, H$_2$ 300 sccm, μwPw: 400 W, T$_s$: 380°C)

7.4 Suppression of Light-Induced Degradation and Improved Performance

Degradation of the a-SiGe:H cells can be suppressed by the reduction of i-layer thickness and by the optimization of the deposition rate. First, the i-layer thickness of bottom and middle cells was reduced from 150 nm to 65 nm. From the reduction of i-layer thickness, the following effects are expected: 1. The internal electric field is increased. 2. Photogenerated carriers need to move a smaller distance to get out of the i-layer. Both contribute to the reduction of carrier recombination in the i-layer, which is thought to be the main cause of photoinduced degradation.

An a-SiGe:H control cell with a 150-nm thick i-layer deposited at 10 nm s^{-1} showed an initial efficiency of 8.2% (V_{oc} = 0.696 V, J_{sc} = 19.99 mA cm^{-2}, FF = 0.589) and a stabilized efficiency of 4.0% (degradation ratio 51%). The cell with the thinner (65-nm thick) i-layer at 10nm/s (the bandgap

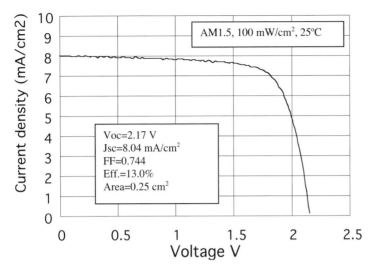

Fig. 7.6. I–V charateristics of a-Si:H/a-Si:H/a-SiGe:H stacked solar cell

was made smaller intentionally) showed a slightly lower initial efficiency of 7.7% ($V_{oc} = 0.64$ V, $J_{sc} = 19.6$ mA cm^{-2}, FF $= 0.616$), but its stabilized efficiency stood at 5.0% (degradation ratio 35%).

Next, the deposition rate of the i-layer was deliberately lowered for better film quality and improved interface characteristics. By doing this, a successful improvement on the stability of the cell was achieved while keeping its initial efficiency value. The improved cell exhibited an initial efficiency of 8.2% ($V_{oc} = 0.646$ V, $J_{sc} = 20.81$ mA cm^{-2}, FF $= 0.608$) and a stabilized efficiency of 5.9% (degradation ratio 28%) [4]. To maintain the device performance with a thinner i-layer, a substantial effort was made to improve the properties of the ZnO/Ag back reflector. An increased random roughness of ZnO surface can make up for the loss caused by the decrease in film thickness.

Through the R&D work, it was found that a lower deposition rate of 4.7 nm s^{-1} for a-SiGe:H deposition on the improved back reflector was useful. The lowered deposition rate is almost half the value initially used, but the thickness of the i-layer is also made half of the former cell. A series of R&D efforts made it possible to produce small-degradation a-SiGe:H solar cells with high throughput.

The developed a-SiGe:H single cells were evaluated for triple-cell uses. To simulate the actual triple-cell performance condition, the degradation test should be done with the short-wavelength cutoff optical filters [5]. For the degradation test of the middle cell, a 570 nm cuton filter was placed between the solar simulator (AM1.5, 100 mW cm^{-2}) and the middle cell. For the bottom cell, a 630 nm cuton filter was used [4]. Other degradation conditions (50°C, open-load) were the same as the degradation test without optical filters. The results of the filtered light degradation test are shown in Fig. 7.7.

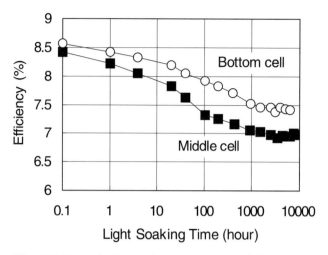

Fig. 7.7. Degradation on characteristics of a-SiGe:H middel cell (with 570-nm cuton filter) and bottom cell (with 630-nm cuton filter)

The initial efficiencies of the middle and bottom cells were 8.4% and 8.6%, respectively. The degradation curves of both middle and bottom cells tend to saturate over 1000 h light exposure. The stabilized efficiencies of middle and bottom cells were 7.0% (degradation ratio 17%) and 7.4% (degradation ratio 13%), respectively. This result demonstrates that the a-SiGe:H cells deposited by LP-MWCVD can have small degradation, while they are deposited at very high rate and at high gas dissociation ratio. The degradation ratio of the middle and bottom cells made at low rate by RF was found to be comparable to the high-rate cell made by the by microwave process, implying that in the deposition rate regions below ~ 4.0 nm s^{-1}, the magnitude of degradation may not be strongly dependent on deposition rate. Degradation tests were conducted on the optimized a-Si:H/a-SiGe:H/a-SiGe:H triple-junction cell by the LP-MWCVD method for a-SiGe:H deposition. The degradation conditions were AM1.5, 100 mW cm^{-2}, 50°C, open-load. The result is shown in Fig. 7.8.

The efficiency of the triple cell (initial efficiency 11.7%) showed a tendency to saturation over 1000 h light exposure. The stabilized efficiency after 21 355 h light soaking was 10.2% with a degradation ratio of 13% [4].

A series of R&D efforts made with the small-area cell on the batch-type CVD apparatus bore fruit by overcoming the technological challenges pointed out in Sect. 7.1 (1. Higher conversion efficiency 2. Higher deposition rate 3. More efficient source-gas utilization).

The fruits of the R&D efforts were applied to the roll-to-roll production process that yields unique integrated solar-cell modules for rooftop uses.

Of course, not all the results we obtained in the small-area cell experiments were used to develop the production machine, because mass production faces technological challenges and limitations of its own. It might be proper to

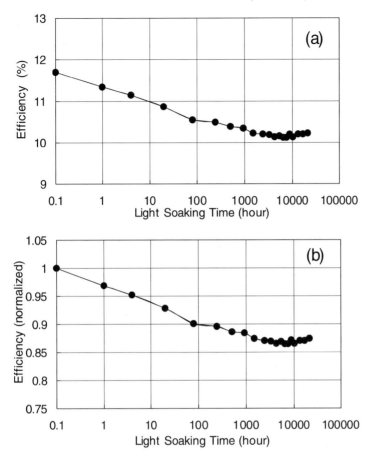

Fig. 7.8. (a,b). Degradation on characteristics of an optimized triple cell

say that our production machine reflects the essence from the small-area cell results.

7.5 Mass Production of a-Si:H/a-SiGe:H/a-SiGe:H Stacked Solar Cells and the Product Outlines

Building-material integrated photovoltaic (PV) modules are generally fabricated on glass or stainless steel substrates. Excellent processability is characteristic of the PV modules using stainless-steel substrate cells, and they are preferably processed into the roofing-material integrated PV modules [6]–[8]. They are already in practical use as "the power-generating roofs" of more than 1000 private houses and public facilities. Figure 7.9 shows an example

Fig. 7.9. Example of the installed roofing-material integrated PV modules

of the installed roofing-material integrated PV modules. It is seen that the
stainless-substrate modules make their appearance as natural as a usual roof.

In this section we describe the fabrication process and the output performance of the building-material integrated PV modules made with the
stainless-steel substrate cells.

The production sequence of the triple-junction cell module is illustrated in
Fig. 7.10. From web washing to ITO deposition, each process has a roll-to-roll

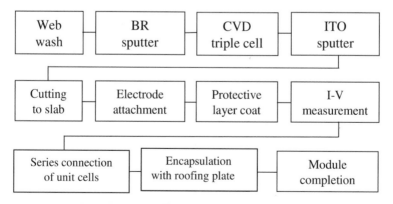

Fig. 7.10. Production sequence of the roofing-material integrated PV modules

apparatus of its own. After the ITO deposition is finished, the web is cut at an interval of 239 mm to make the slab cell. The integrated modules provided for rooftop applications are made by encapsulating the whole combined slab cells (or unit-size cells). Detailed descriptions of each production step are as follows:

Web Washing

The 356-mm wide stainless roll web (active deposition length 820 m) is first washed in a detergent bath to make its surface clean and suitable for thin-film deposition.

Back-Reflector Deposition

The washed roll is then set on the sputtering machine to form a textured ZnO/Al back reflector.

Formation of the Triple-Junction Cell

The roll-to-roll CVD machine is used to deposit the a-Si:H/a-SiGe:H/a-SiGe:H triple-junction cell. The knowledge from the basic work in Sect. 7.1 was fully utilized for the deposition of bottom and middle a-SiGe:H i-layers by the microwave process at a considerably higher rate on the order of several nanometers per second. High rate deposition of the i-layers by the MW-PCVD method makes it possible to produce high-performance a-Si:H/a-SiGe:H/a-SiGe:H triple-junction cells at a relatively high web speed.

ITO Deposition

For ITO deposition, the reactive sputtering method was used with an indium-tin target in an oxygen/argon gas mixture. This method can reduce production cost because the indium-tin target is available at a far lower cost than the sintered ITO target.

Web Cutting

The ITO-deposited web was cut at a constant interval to make the 356×239 mm size slab cells.

Metal Electrode Attachment

Current collection electrodes consisting of metal wires and busbars were attached onto the ITO surface to complete the slab-cell fabrication.

Coating of Protective Layer

The surface of the electrode-attached slab cell was coated with the hard protective layer.

I–V Measurement of the Slab Cell

For quality assurance purposes, the performance of each slab cell is monitored by measuring the I–V characteristics under 1 sun white light.

Unit-Size Cell Preparation

The good slab cells are cut into smaller-size unit cells, if necessary. Regular-size (356×239 mm) slab cells are also used as a unit-size cell.

Encapsulation

Unit cells are combined and encapsulated. In the process depicted in Fig. 7.10, various types of unit-size cells are combined to fabricate the integrated modules. This choice allows one to provide integrated modules of various sizes and shapes according to the needs of customers.

7.6 The Roll-to-Roll CVD Method

The roll-to-roll CVD process is believed to be one of the most promising production methods for thin-film solar cells. Canon have chosen this method to mass produce a-Si:H/a-SiGe:H/a-SiGe:H triple-junction stacked solar cells mentioned in Sect. 7.1. The schematic of the roll-to-roll CVD apparatus (for semiconductor layer deposition) is shown in Fig. 7.11.

The stainless roll web substrate is mounted on the pay-off chamber, and the nip/nip/nip semiconductor layers are deposited successively while the web moves continuously through the adjacent chambers. A photograph of the CVD production machine is given in Fig. 7.12.

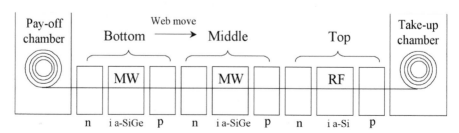

Fig. 7.11. Schematic of the roll-to-roll CVD apparatus

Fig. 7.12. Roll-to-roll CVD apparatus in operation

The series-connected cells in Fig. 7.10 are encapsulated with a roofing steel plate to make a solar-cell module. The edge portions of the plate can be bent easily with the roller former or bending machine. Therefore, it can respond to various demands in designing and can be used for various kinds of metal roofs. The bending processing also contributes to giving a high rigidity to the module, which satisfies the installation requirements (shape, strength, etc.) without additional parts such as an aluminum frame.

A few examples of the roofing modules are given in Fig. 7.13a through Fig. 7.13c. The model in Fig. 7.13a is called the "batten-and-seam roof", a type with its edge portions bent upright. As shown in Fig. 7.13a, the modules are placed on the stiffeners and fixed with clip tingles or capping components. Simplicity in processing and installation is the characteristic feature of this model.

The model in Fig. 7.13b is called the "stepped roof". The edge portions of this model are processed to form hook-like shapes. The modules can make the entire roof by only connecting the adjacent modules with the hook-like shapes. The completed roof appears nice and natural. Excellence in design quality is characteristic of this model.

The model in Fig. 7.13c is called the "flat roof". The completed module length ranges from 2.1 to 5.4 m. The edge portions are bent so as to fit into the fastening part. The modules edges are placed into the support body called the gutter and sealed with the capping component to complete the roofing fabrication. Installation of this module is quite easy. When used on a

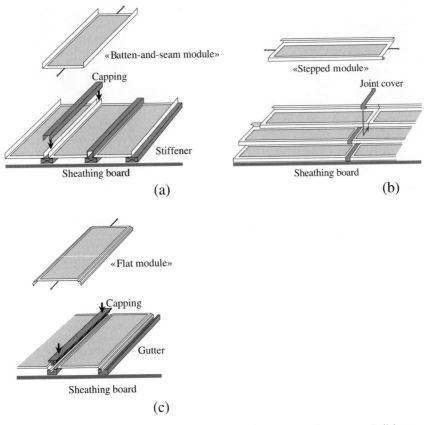

Fig. 7.13. Examples of the roofing modules; (**a**) batten-and-seam roof, (**b**) stepped roof, (**c**) flatroof

large-size building, this model can make the appearance of the building quite neat.

7.7 Characteristics of Slab Cells and Modules

Initial Performance

Initial performance of each slab cell is examined for quality control purposes. The I–V measurement is done at a cell temperature of 25°C under an AM1.5, 100 mW cm^{-2}, white light. Variation of the initial performance of approximately 3600 slab cells through the 820 m long roll web is shown in Fig. 7.14. Also, the initial-efficiency distribution of the 3600 slab cells is depicted in Fig. 7.15.

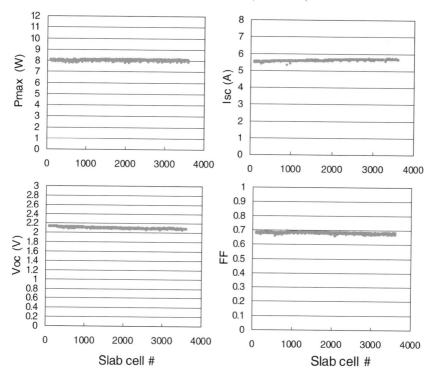

Fig. 7.14. Performance of slab cells through the 820 m long roll web

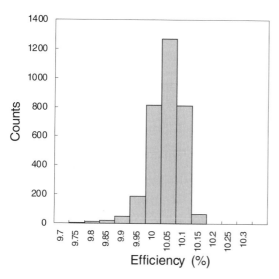

Fig. 7.15. Distribution of initial efficiencies in % through the 820 m long roll web

It is confirmed that the well-developed roll-to-roll process yields a considerably smaller performance variation through the entire roll length; the initial output power (initial efficiency), V_{oc}, I_{sc}, and FF were in the ranges of 7.980 ± 0.180 W ($9.938 \pm 0.224\%$), 2.101 ± 0.042 V, 5.54 ± 0.21 A, and 0.678 ± 0.019, respectively. The average initial values of these parameters were 8.044 W (10.02%), 2.103 V, 5.620 A, 0.680, respectively. These data prove that the roll-to-roll process can be effectively applied to the mass production of solar cells.

Degradation Characteristics

From a series of degradation tests on the slab cell, it was found that the degradation ratio of the mass-produced slab cell was not so different from that obtained from the small-area cell. Through the continuing effort for better stability, slab cells with almost 9% stabilized efficiency have been obtained. Considering the fact that these cells are produced at a high deposition rate of several nanometers per second, this stabilized efficiency is worth noting.

7.8 Light-Soaking Testing

Presently, an initial efficiency of about 10% has been achieved in the rooftop module as a commercial product. The intrinsic average initial efficiency of the component cells is about 11%. There is an efficiency gap of about 9%. The loss factors in the mentioned module configuration were summed by taking into account the electrode shadows, electrical resistance, and optical losses by the encapsulation material. The calculated value was almost the same as the above mentioned efficiency gap. An initial efficiency of 13.0% has been reported [4] in a small-area cell deposited at high rate with the laboratory-size, batch-type CVD apparatus. The optimized high-rate, small-area cell (with the highest stabilized efficiency) exhibited an initial efficiency of 11.7% [4]. These data indicate that the performance level of our production modules is as high as that of the optimized laboratory-size cell.

Figure 7.16 shows a comparison of the degradation characteristics of the production cell and the laboratory-size cell. The degradation pattern of the production cell is similar to that of the laboratory cell. Both curves start to saturate around 1000 h light soaking and they become almost stable beyond 1000 h. This means that the degradation characteristics of the production cell have been improved to a level comparable to that of the laboratory cell. Since the component cells of the mentioned modules are all made with a-Si based materials, they can enjoy performance recovery at higher working temperatures by the thermal annealing effect. It is confirmed that the a-Si:H/a-SiGe:H/a-SiGe:H module, contrary to the c-Si counterpart, can work more efficiently in high annual-average-temperature regions [9].

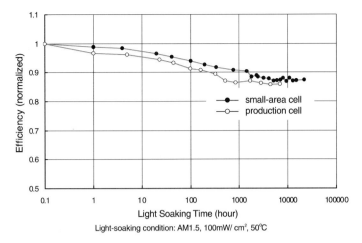

Fig. 7.16. Comparison of the degradation on characteristics of small-area (laboratory-size) cells and production size cell (light-soaking condition: AM 1.5, 100 mW cm^{-2}, 50°C)

7.9 Summary

R&D efforts have been made to obtain high-efficiency a-Si:H/a-SiGe:H/a-SiGe:H stacked solar cells and module production for rooftop applications. The low-pressure microwave plasma chemical vapor deposition (LP-MWPCVD) method combined with the roll-to-roll process made it possible to mass produce high-efficiency triple-junction cells at high throughput. The intrinsic cell efficiency of the roll-to-roll production cell has been improved to the level achieved in the laboratory-size cell. The stabilized efficiency of the latest mass-produced module is improved to almost 9%. The modules with the stainless-steel substrate cells can be regarded as a metal roofing material. Installation simplicity and beautiful appearance are characteristic of these roofing modules. They are already applied to various types of roofs in residential houses and public facilities.

References

1. Hamakawa, Y., and Kuwano, K. (1994): "Solar Energy Engineering", Baihu-kan, 5, in Japanese.
2. Saito, K., Sano, M., Ogawa, K., and Kajita, I. (1993): "High Efficiency a-Si:H Alloy Cell Deposited at High Deposition Rate", *J. Non-Cryst. Solids* 164&166, 689.
3. Saito, K., Sano, M., Matsuyama, J., Higasikawa, M., Ogawa, K., and Kajita, I. (1996): "The Light Induced Degradation of the a-Si:H Alloy Cell Deposited by the Low Pressure Microwave PCVD at High Deposition Rate", Tech. Digest 9th Int. PVSEC, Miyazaki, Japan, 579.

4. Sano, M., and Saito, K. (1999): "A-Si Based Solar Cells Deposited at High Rate by the Microwave PCVD Method", Textbook 26th Seminar on Amorphous Materials, Hakata, Japan, 33, in Japanese.

5. Guha, S., Yang, J., Banerjee, A., Glatfelter, T., Hoffman, K., and Xu, X. (1993): "Progress in Multijunction Amorphous Silicon Alloy-based Solar Cells and Modules", Tech. Digest 7th Int. PVSEC, Nagoya, Japan, 43.

6. Kajita, I. (1996): "The Residential Photovoltaic Power Generation", Proc. 13th Photovoltaic Power Generation System Symposium.

7. Fukae, K., Mori, M., Itoyama, S., Inoue, Y., Mori, T., Toma, H., and Kajita, I. (1996): "Photovoltaic Cell Integrated Roofing Module", Proc. 14th International Conference on Passive and Low Energy Architecture, Kushiro, Japan, Vol. 1.

8. Takeuchi, E., and Fukae, K. (1994): "Applications of a-Si Based Solar Cells – the Roofing-material Integrated PV Module", Kenchiku-Setsubi (The Magazine of Building Equipment) 518, No.5, 40, in Japanese.

9. Fukae, K., Lim, C.C., Tamechika, M., Takehara, N., Saito, K., Kajita, I., and Kondo, E. (1996): "Outdoor Performance of Triple Stacked a-Si Photovoltaic Module in Various Geographical Locations and Climates", Proc. 25th IEEE-PVSC, Washington, D.C., 1227.

Low-Temperature Fabrication
of Nanocrystalline-Silicon Solar Cells

Michio Kondo and Akihisa Matsuda

A novel approach of the low-temperature processing for nanocrystalline silicon is overviewed. Contrary to the empirical rule between the solar-cell efficiency and the crystallite size, nanocrystalline-silicon-based solar cells demonstrate an efficiency higher than 10% in spite of the small size of 10 nm which is several orders of magnitude smaller than the conventional polycrystalline materials with a comparable efficiency. A predominant cause of the high efficiency arises from hydrogen, which passivates the grain-boundary defects. The hydrogen has other beneficial roles such as improvement of the crystalline fraction at lower temperatures, improvement of the crystallinity for boron-doped thin layers as used windows, and passivation of oxygen-related donors that decrease the open-circuit voltage. These beneficial roles overcome the disadvantage of small crystallite size at low-temperatures. Nanocrystalline-silicon solar cells on plastic substrate are also demonstrated.

8.1 Why Nanocrystalline-Silicon Solar Cells?

The most popular solar cells at present are made of single- and multicrystalline silicon. The single-crystalline silicon and multicrystalline (or "conventional" polycrystalline) silicon solar cells show efficiencies above 24% and 17%, respectively [1]. The lower efficiency of multicrystalline silicon is ascribed to the carrier recombination at the grain-boundary defects, and therefore one can find a good correlation between the efficiency, particularly open-circuit voltage (V_{oc}) and the grain size, as shown in Fig. 8.1 [1]. The carrier recombination dominates the saturation current and thereby affects the open-circuit voltage. The recombination occurs at dangling-bond defects in the grain boundaries. Simple extrapolation of the correlation between the V_{oc} and the grain size indicates quite low V_{oc} as well as efficiency for the grain size of 10 nm that

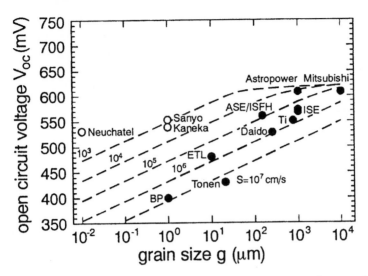

Fig. 8.1. Empirical relation between the open-circuit voltage and the grain size for polycrystalline silicon (closed circle) and micro- (nano-)-crystalline silicon (open circles) (after [1])

is a typical grain size of nanocrystalline silicon. Thus, no one believed in the possibility of nanocrystalline-silicon solar cells for practical application.

However, the recent development of nanocrystalline silicon solar cells reveals the possibility of high efficiency in spite of the small grain size of the order of 10–100 nm [2, 3]. Efficiencies higher than 10% have been obtained at process temperatures less than 400°C [4]. As shown in Fig. 8.1, nanocrystalline silicon shows much higher V_{oc} than extrapolated values from the conventional polycrystalline-silicon solar cells. It is astonishing that a difference in the grain size by several orders of magnitude leads to a difference in the V_{oc} by only 10%. This high efficiency has been ascribed to the hydrogen passivation of the grain-boundary defects and the suppression of the recombination rate at the grain boundaries. Nanocrystalline silicon is commonly fabricated at temperatures below 400°C and typically around 200°C, as mentioned in Chap. 4. This low-temperature process reduces the grain size, whereas a large amount of hydrogen, of the order of several atomic per cent, remains in the film and passivates the dangling-bond defect at the grain boundaries as in hydrogenated amorphous silicon [5]. The defect density of the nanocrystalline silicon fabricated at around 200°C is below 10^{16} cm^{-3}, as measured by electron spin resonance (ESR) [6]. This is consistent with the expected low recombination rate in solar cells as shown in Fig. 8.1. Thus, it has been demonstrated that nanocrystalline silicon possibly attains an efficiency (> 10%) comparable to conventional polycrystalline-silicon solar cells thanks to the hydrogen passivation of the grain boundaries.

Another advantage of nanocrystalline silicon is a possibility of higher efficiency by means of the tandem structure in combination with an a-Si:H solar cell [2, 7]. The low process temperature enables us to fabricate either superstrate- or substrate-type structure on inexpensive substrates such as glass, polymer, and stainless steel. Thus, the nanocrystalline silicon fabricated at low-temperatures is a promising material for thin-film solar cells for the next generation.

8.2 Low-Temperature Process
for Nanocrystalline-Silicon Solar Cells

Nanocrystalline-silicon solar cells are commonly operated in a pin device structure because of the short carrier diffusion length, and the fac that photocarriers are driven by the built-in field. Since the i-layer contains defects of the order of 10^{15-16} cm^{-3}, the band bending at the p/i and n/i interfaces causes the upper limit of the i-layer thickness for the extraction of the carriers. However, some experimental results suggest the carrier diffusion length over several hundreds nm, which is much larger than the grain size evaluated from XRD. The cross-sectional TEM image reveals the anisotropy with columnar shape and the diffusion length might be dominated by the longer dimension along the z-axis of the column. In fact, anisotropic carrier transport has been observed [8].

In solar cells, doped layers are crucial for device performance. The p-layer acts as a window layer, and therefore not only the electrical properties but also the optical transmittance in the short-wavelength region are of great importance. The nanocrystalline-silicon p-layer has advantages over amorphous silicon and amorphous silicon carbide in the respects of higher electrical conductivity and lower optical absorption. The plasma-enhanced chemical vapor deposition (PECVD) has been most widely used for the deposition of p, i, and n layers. The p and n layers are formed by adding diborane and phosphine gases as dopants, respectively. The presence of the impurity influences the crystal growth. Undoped films shows the best crystallinity around 400°C, as explained in Chap. 4, while B-doped films do so at around 140–180°C as shown in Fig 8.2. This difference is ascribed to the catalytic role of boron to eliminate surface hydrogen [9]. This result also supports the importance of the surface coverage by hydrogen.

Therefore, it turns out that the low-temperature deposition of the p-layer is effective to improve the crystallinity. In addition, low-temperature deposition is beneficial to prevent the bandgap narrowing of the B-doped amorphous tissue region and the darkening of the TCO substrate due to the reduction by atomic hydrogen [10]. The electronic conductivity, however, is lowered due to the hydrogen passivation of dopants [11], and it has been reported that the post-annealing treatment is effective to improve the conductivity [12]. Another important requirement is the optical properties. In order to reduce

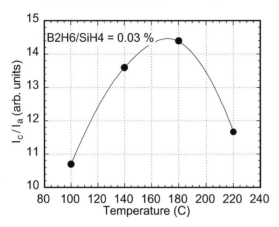

Fig. 8.2. Deposition-temperature dependence of the Raman crystallinity, I_c/I_a, for B-doped films. For undoped films, see Fig. 4.4 for comparison.

the optical absorption in the doped layers, the thickness should be minimized, while a certain amount of the dopants is required to form the built-in field, because there are dangling bonds that can act either as a donor or acceptor in the i-layer. Therefore, the doped layers should be heavily doped, while the p-layer is difficult to make crystalline as demonstrated in Fig. 8.3. The conductivity reveals a crystallinity and fully crystallized films usually show a conductivity over 0.1 S cm^{-1}. Figure 8.3 implies that there is a difficulty during the crystalline nucleation for high boron concentration for thickness below 200 Å, which is a typical thickness of the p-layer.

In the i-layer, the major impurities are oxygen, carbon and nitrogen. The oxygen-related donors, if activated, act as space charge and thereby screen the built-in potential in the i-layer and the stronger field is formed at the p/i interface as compared to the "intrinsic" i-layer. This results in the reduction in the "active" layer thickness in the i-layer. It has been reported that the spectral response of the contaminated cell shows a marked deterioration in the long-wavelength region. This implies that the photoexcited carrier (holes) cannot be extracted to the p-layer due to the weakened field ("dead" layer) in the deeper part of the i-layer. Even with oxygen contamination, it has been found that the low-temperature deposition successfully passivates the oxygen-related donors as mentioned in Chap. 4. The deposition temperature, however, is a very complicated parameter because it affects a variety of material properties such as crystallinity, defect density, interface damage, and so on. In the case of the usual vacuum level ($\sim 10^{-5}$ Pa), the incorporated oxygen is $> 10^{19}$ cm^{-3} and the oxygen-related donors have been deactivated by reducing the deposition temperature to 140°C. This low-temperature deposition results in a markedly improved V_{oc}, while the J_{sc} is still maintained at a reasonable level, as shown in Fig. 8.4. As a consequence, the optimum temperature is 140–180°C depending on other device parameters, such as the

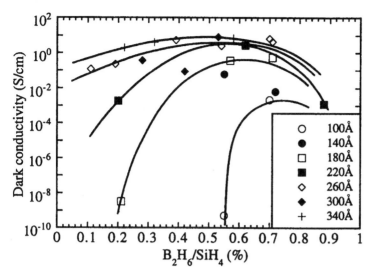

Fig. 8.3. Thickness dependence of the room-temperature conductivity for B-dope layers with various thickness (after [13])

i-layer thickness, and an efficiency of 8.6% was obtained on the Asahi-U substrate with a 2.6-μm thick i-layer, and 9.4% on the textured ZnO:Ga substrate with the improved texture as mentioned later [14]. It is still an open question, however, whether a further improvement in efficiency is possible at higher temperature without oxygen contamination because the defect density of undoped nanocrystalline silicon shows a minimum at $\sim 200°C$ [7] similarly to a-SI:H.

Under the fixed deposition conditions, the crystalline volume fraction decreases with decreasing temperature as shown in Fig. 4.4, and the cell efficiency also decreases, as shown in Fig. 8.4. This can be recovered by increasing the hydrogen dilution ratio. Under variable hydrogen dilution conditions for optimizing the efficiency, it turns out that the efficiency can be maintained down to 120°C, as shown in Fig. 8.5. In this temperature regime, the hydrogen passivation of the dopants becomes a more serious problem. At present, the post-annealing process is the only way to overcome this problem. The upper limit of the processing temperature, on the other hand, seems to be determined by the thermal damage at the p/i and n/i interfaces. The nip cell can be successfully obtained over 300°C [4], while the pin cell is usually prepared at $\sim 200°C$ or the lower [2, 16] . These results could arise from the different thermodynamics of the dopants.

Hydrogen dilution is an essential factor to determine the solar-cell performance as well as the crystallinity. The hydrogen-dilution dependence of the V_{oc} is shown in Fig. 8.6, [15]. With increasing hydrogen dilution, the structure varies from amorphous to crystalline and the crystalline volume fraction

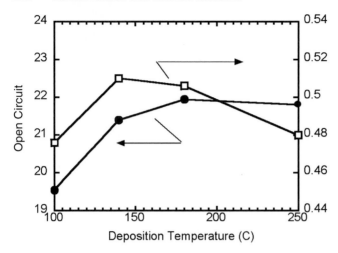

Fig. 8.4. Deposition-temperature dependence of V_{oc} and J_{sc}. The substrate is Asahi-U type. The i-layer thickness is 1 μm (after [12])

increases monotonically. Therefore, one may expect the better performance for higher dilution ratio, whereas this is not the case. V_{oc} is monotonically increases with decreasing dilution ratio. The J_{sc}, on the other hand, is nearly constant in the crystalline regime and decreases abruptly in the amorphous regime. Therefore, an empirical conclusion is that the best performance is obtained near the amorphous/crystalline phase boundary in the crystalline regime.

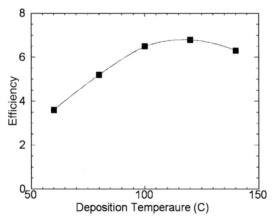

Fig. 8.5. Dependence of the efficiency optimized at each temperature on the i-layer deposition temperature. The i-layer thickness is fixed at 2 mm and the flat substrate was used [12]

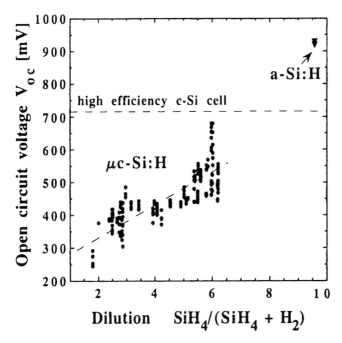

Fig. 8.6. Hydrogen-dilution-ratio-dependence of the open-circuit voltage for pin solar cells (after [15])

8.3 Substrate Technology

In amorphous-silicon solar cells, textured substrates are used to reduce the i-layer thickness on the basis of the light-trapping technique, because the weakened built-in field in the thick i-layer results in significant photo-degradation. In nanocrystalline-silicon solar cells, on the other hand, the required thickness for the i-layer is at least 1 μm because of its indirect bandgap. Taking into account the typical defect density of 10^{16} cm^{-3}, the band bending (or screening of the built-in field) due to the charged defect results in a poor fill factor and open-circuit voltage for the i-layer thicker than several mm. Therefore, in nanocrystalline-silicon solar cells, the light-trapping technique is of great importance, particularly in the long-wavelength region ($\lambda > 700$ nm) as compared to the a-Si:H solar cells. For pin-type amorphous-silicon solar cells, textured SnO$_2$ is widely used, and has been also applied for nanocrystalline-Si solar cells [15, 16]. However, the optimized texture is not always the same for nanocrystalline-silicon solar cells, and therefore different approaches using ZnO have been attempted by means of wet etching [17] and LPCVD [18].

An additional issue, other than the optical effect, emerges in the case of nanocrystalline-silicon solar cells. Figures 8.7 shows the cross-sectional TEM image of μc-Si layers on the three different textured substrates; (a) flat ZnO, (b) textured ZnO fabricated by the wet-etching technique, and (c) Asahi-

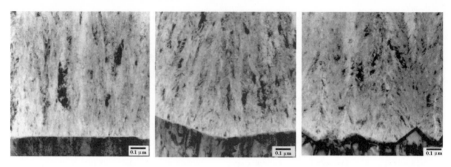

Fig. 8.7. (a-c). Cross-sectional TEM images of nanocrystalline silicon on different surface texture (after [16])

U coated with 500-Å thick ZnO. The overcoating of the ZnO is effective to reduce the darkening of the SnO_2 due to the reduction by atomic hydrogen [10]. Looking at the growth trajectory of nc-Si layer, the columnar growth along [110] direction takes place, and the growth direction is perpendicular to the local surface as indicated by the arrow in the figure. On the textured surface, the columnar structure starting from different local surfaces collides with each other and a macroscopic grain boundary is formed near the p/i interface, and correspondingly, the V_{oc} decreases from (a) to (c).

This result therefore suggests that the substrate texture optimized for nc-Si:H solar cells is different for a-Si:H solar cells. In the case of the nip structures, the STAR structure was successfully introduced on the flat substrate utilizing the spontaneous texture of the crystal growth [3]. For the pin structure, the optimized structure is rather shallower than the Asahi-U as demonstrated in Fig. 8.7b and 8.7c. The best efficiency on Ashai-U (Fig. 8.7c) has been reported to be 8.5–8.6% [15, 16], while on the shallower texture (Fig. 8.7b) an efficiency of 9.4% has been reported [13, 16], as evidenced in Fig. 8.8.

For the cost reduction of solar cells, the roll-to-roll process has been considered as a promising method, and for this purpose, flexible sheet substrates are required. Stainless steel and polymer substrates have been developed for amorphous-silicon solar cells [19, 20]. The polymer substrate is attractive for various uses such as mobile electricity sources and biomechanical applications. However, the common drawback of polymers is durability at high temperatures. Commodity plastic has a low glass-transition temperature that limits the application of the polymer substrates to solar cells. The recent progress, however, in high-temperature durability of thermoplastic and in low-temperature processing techniques for nanocrystalline silicon, as mentioned above, enables us to apply these materials to nanocrystalline-silicon solar cells. The actual application on the plastic substrate has been demonstrated by several groups using E/TD (ethylene tetracyclododesence copolymer). Figure 8.9 shows a picture of the microcrystalline-silicon solar cell on

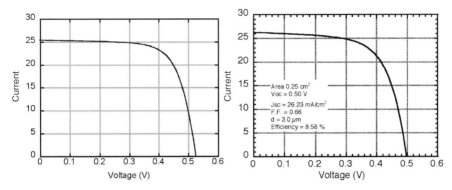

Fig. 8.8. I–V characteristics of nanocrystalline-silicon solar cell with an efficiency of 9.4% (left) and 8.6% (right) on the substrate (b) and (c), respectively, as showing in Fig. 8.7 (after [13, 16])

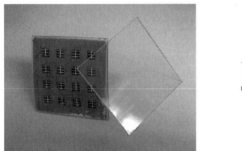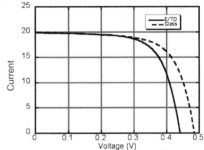

Fig. 8.9. Example of the nanocrystalline-silicon solar cell on the polymer substrate and its I–V characteristics. The I–V curve for the cell on the glass substrate prepared under the identical conditions are shown for comparison (after [21])

the E/TD substrate and its cell performance. The deposition temperatures for all the layers was kept at 100°C and all the process temperature including the ITO deposition and annealing procedures. The efficiency of 6% is reasonable, taking into account the nontextured substrate. The identical process was applied to the glass substrate and an efficiency of 7% was obtained as shown in Fig. 8.9 [21].

8.4 Future Prospect of Nanocrystalline-Silicon Solar Cells

In summary, we have argued the advantages and disadvantages of nanocrystalline-silicon solar cells from a variety of viewpoints such as fundamental growth, material properties (Chap. 4), and device performance. Several years ago, no one believed that tiny crystallites smaller than 1/100 of the conventional polysilicon could be used as actual devices with a promising efficiency over 10%. Simple thermodynamic considerations support the very low growth rate and poor crystallinity at low-temperatures. In fact, at that time, the efficiency of the nanocrystalline-silicon solar cells was very low compared to the amorphous-silicon counterpart. Recent progress, however, in the material as well as the devices of nanocrystalline silicon has unveiled its potential performance, and there seems no reason to insist on the advantage of the high-temperature polysilicon unless the device performance is much better than 10%. It is also noteworthy that these improvements are based on the beneficial role of hydrogen and that the hydrogen has been considered to disrupt the crystal growth in conventional CVD and MBE. For further development, we should overcome the serious drawback of nanocrystalline silicon that is, the lower open-circuit voltage than high-temperature polysilicon. There has been an argument concerning the upper limit of the V_{oc} being 550 mV for nanocrystalline silicon [22], and in fact, the current status of the maximum V_{oc} is around 550 mV. Another possibility of nanocrystalline silicon is to utilize the quantum-size effect to obtain higher V_{oc} due to the larger (quasi-)direct bandgap.

Acknowledgments

Finally, the authors would like to acknowledge fruitful collaboration with Drs. Y. Nasuno, H. Yamamoto, S. Suzuki, T. Wada, H. Mase, T. Yamamoto, M. Fukawa, L. Guo, K. Saitoh, and H. Fujiwara.

References

1. Bergmann, R.B. (1999): *Appl. Phys.* A 69, p.155.
2. Meier, J., et al. (1996): *Proc. Mater. Res. Soc. Symp.*, 420 p.3.
3. Yamamoto, K., Suzuki, T., Yoshimi, M., and Nakajima, A. (1997): *Jpn. J. Appl. Phys.*, 36 L569.
4. Yamamoto, K., Yoshimi, M., Tawada, Y., Okamoto, K., Nakajima, A., and Igari, S. (1999): *Appl. Phys.* A, 69 p.179 and references therein.
5. Spear, W.E. and LeComber, P.G. (1975): *Solid State Commun.*, 17 p.1193.
6. Kamei, T., Kondo, M., and Matsuda, A. (1998): *Jpn. J. Appl. Phys.* 37 p.265.
7. Hamakawa, Y., Okamoto, H., and Nitta, Y. (1979): *Appl. Phys. Lett.*, 35 p.187. US pat. 4,271,328, Jan. 2, 1981.
8. Fejfar, A., Beck, N., Stuchlikova, H., Wyrsch, N., Torres, P., Meier, J., Shah, A., and Kocka, J. (1998): *J. Non-Cryst. Solids*, 227–230, p.1006.
9. Perrin, J., Takeda, Y., Hirano, N., Takeuchi, Y., and Matsuda, A. (1989): *Surf. Sci.*, 210 p.114.
10. Ikeda, T., Sato, K., Hayashi, Y., Wakayama, Y., Adachi, K. and Nishimura, H. (1994): *Solar Energy Mater. Solar Cells*, 34 p.379.
11. Pankov, J.I., Zanzucchi, P.J., Magee, C.W., and Locpvsky, G. (1985): *Appl. Phys. Lett.*, 46 p.421.

12. Kondo, M., Nasuno, Y., Mase, H., and Matsuda, A. (2002): *J. Non-Cryst. Solids*, in press.
13. Fluckiger, R., Meier, J., Shah, A., Catana, A., Brunel, M., Nguyen, H.V., Collins, R.W., and Carius, R. (1994): *Proc. Mater. Res. Soc. Symp.* 336, p.511.
14. Nasuno,Y., Kondo, M., and Matsuda, A. (2001): *Jpn. J. Appl. Phys.*, 40 p.303.
15. Meier, J., Keppner, H., Dubail, S., Kroll, U., Torres, P., Pernet, P., Ziegler, Y., Anna Selvan, J.A., Cuperus, J., Fischer, D., and Shah, A. (1999): *Proc. Mater. Res. Soc. Symp.*, 507 p.139.
16. Nasuno, Y., Kondo, M., and Matsuda, A. (2001): *Appl. Phys. Lett.*, 78 p.2330.
17. Kluth, O., Rech, B., Houben, L., Wieder, S., Schope, G., Beneking, C., Wagner, H., Loeffl, A., and SchoCk, H.-W. (1999): *Thin Solid Films*, 351 p.247.
18. Meier, J., et al. (Jeju, 2001): Tech Digest PVSEC12 p.783.
19. Ichikawa, Y., Tabuchi, K., Takano, A., Fujikake, S., Yoshida, T., and Sakai, H. (1996): *J. Non-Cryst. Solids*, 198-200 p.1081.
20. Guha, S., Yang, J., Banerjee, A., Glatfelter, T., and Sugiyama, S. (1997): *Sol. Energy Mater. Sol. Cells*, 48 p.365.
21. Mase, H., Kondo, M., and Matsuda, A. (Jeju, 2001): Tech Digest PVSEC12, p. 473.
22. Werner, J.H., Dassow, R., Rinke, T.J., Kohler, J.R., and Bergmann, R.B. (2001): *Thin Solid Films*, 383, p.95.
23. Keppner, H., Meier, J., Torres, P., Fischer, D., and Shah, A. (1999): *Appl. Phys. A*, 69 p.169 and references therein.

Mass Production of Large-Area Integrated Thin-Film Silicon Solar-Cell Module

Yoshihisa Tawada, H. Yamagishi, and K. Yamamoto

A mass-production technology of a-Si single junction modules with stable 8% efficiency had been developed in the Shiga factory of the Kaneka Corporation. In 1999, Kaneka instituted Kaneka Solartech Corporation (KST) as a subsidiary company and invested in a 20-MW production plant in Toyooka City. There are fully automatic thin-film fabrication equipment and laser scribers equipped with conveyors and the robots installed. The main size of glass substrates used for these modules is $910 \times 910 \times 5$ mm, and the recent production yields constantly exceed 95%.

To improve the module efficiency and decrease the module cost, we proposed the new type of cell structure (hybrid cell) with a-Si and thin-film poly-crystalline Si layers. R&D for the new thin-film hybrid solar device has been accelerated and the stable 10% module technology of the hybrid cells has been completed in the Shiga factory. The stable conversion efficiency of this hybrid cell with 1 cm^2 area showed 12% efficiency.

Through many efforts to enlarge the glow-discharge deposition area in the preparation of the thin-film poly-Si, Kaneka established a new technology to fabricate poly-Si films with 1 m^2 area as well as a-Si films.

Since the hybrid modules can be integrated by the laser-scribing method, we could easily modify the production line of a-Si modules to that of hybrid modules. By adding the glow-discharge deposition equipment for thin crystalline Si to the a-Si solar-cell production line of KST, Kaneka has just begun to produce stable 10% hybrid modules, since April 2001, in Toyooka City.

In this chapter, a series of production technologies of a-Si, thin c-Si and solar cells, performances of modules, applications to the rooftop PV systems will be presented. We estimated the production cost of a-Si solar modules and a-Si/thin c-Si hybrid solar modules. We will also discuss the future business plan of our new type of solar modules and our production lines.

9.1 Introduction

For the commercial applications for consumer use, such as calculators and watches, the PV devices and their industry had been successfully by 1980 [1]. Although the trials to realize the outdoor applications of the stable 6% efficient modules had also been carried out with the production scale of 3 to 5 MW/y, it appeared that these production scales were too small to reduce the module cost to be less than $2/Wp. To achieve this cost reduction, at least a stable 8% efficient module-production technologies with the minimum production scale of more than 40 MW/y must be developed.

Several types of thin-film silicon solar cells are investigated as the candidates of outdoor photovoltaic modules that can be applied to a 100-MW scale of a mass-production line. One of the typical processes is roll-to-roll for flexible-type solar cells proposed by several companies [2, 3]. The other is the inline apparatus for the rigid-type substrates, such as glass. In this chapter, we will introduce a mass-production technology of amorphous-silicon (a-Si) and thin-film polycrystalline-silicon solar cells on large-area glass substrates developed by Kaneka Corporation.

9.2 Performance and Production of a-Si Modules

Since 1993 we have built a pilot plant for a-Si PV modules and developed the mass-production technology for modules with a stable 8% efficiency in the Shiga factory of the Kaneka Corporation [4]. In this pilot plant we have investigated individual aspects for a 20-MW scale of mass-production using large-scale apparatus such as CVD, sputtering, and laser-scribing machines.

More popular device structures are pin double- and triple-junction with a-Si and a-SiGe alloy layers to obtain a high cell performance and reduce the cell degradation caused by the light exposure. However, our selection is a pin single-junction scheme. Our R&D strategies of a-Si photovoltaic modules were determined in 1994 as follows:

Materials: amorphous silicon
⟵ Environmental problem, resources, safety
Structure: single junction
⟵ Simple structure, reproducibility, productivity
Efficiency: Stable 8% Scale: 2 lines of 20 MW/Y
Deadline for mass production: 1999.

To reduce the light-induced changes in the cell performance, the i-layer thickness has been limited below 300 nm. To improve the stabilized efficiency of cells with the i-layer thickness less than 300 nm, we made efforts to obtain a high uniformity of thickness and quality using a new a-Si deposition system with 1 m^2 plasma areas.

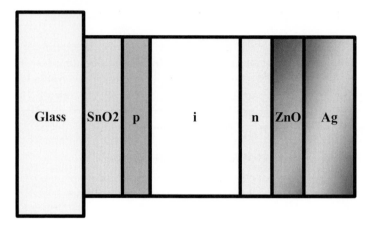

Fig. 9.1. Schematic illustration of amorphous single-junction solar cell produced by Kaneka Solartech Corporation

Figure 9.1 shows a typical cross-sectional structure of the single-junction PV devices produced in the Shiga pilot plant and the Kaneka Solartech Corporation (KST) mass-production lines. To obtain a high short-circuit current density and improve the reliability for the long-term outdoor use, the optical confinement structure is applied between the top TCO and the back-contact metal layers.

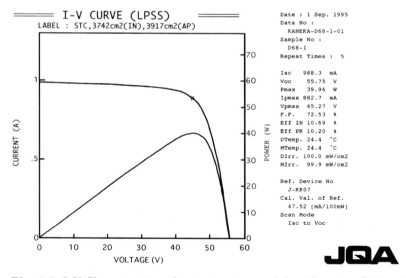

Fig. 9.2. I–V Characteristics of a pin junction module with a size of 910×415 mm. The conversion efficiency, short-circuit current, open-circuit voltage, and fill factor are 10.6%, 1.403 A, 44.3 V, and 67.5%, respectively

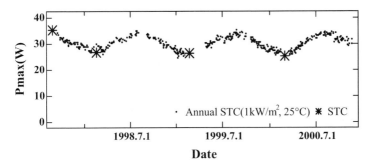

Fig. 9.3. Change of output power by the field test of pin single-junction module in Hamanako by JQA

Figure 9.2 shows the I–V characteristics of a 91 × 45.5 cm module measured by the Japan Quality Assurance Organization (JQA). We have developed a technology to fabricate modules with an initial efficiency more than 10% and the production yield has been improved to be higher than 95% [5].

Figure 9.3 shows the result of the outdoor exposure test for our a-Si module situated at the shores of Lake Hamanako, Shizuoka, Japan by JQA. The module performance changes to 77% of the initial value in winter and recovers to 93% in summer, and shows the reciprocal annual change each year. These data show that the average module performance is estimated at 82% of the initial value. We have not measured the irreversible degradation since 1997. JQA carried out the same field test in Australia and obtained similar results. Through these field data we have confirmed that our modules can be applied to rooftop PV systems.

The investment for massproduction of the a-Si modules was decided with an initial capacity of 20 MW/y in Toyooka City in September 1998. Toyooka City is located in the northern part of Hyogo Prefecture and within 100 miles of the Kaneka Shiga factory. As a subsidiary company of Kaneka, Kaneka Solartech Corporation was established in September 1998 and started to produce and supply the standard a-Si modules in October 1999 [6]. The submodule production lines in KST are fully automatically, operated with the conveyors and the robots as seen in Fig. 9.4.

Figure 9.5 shows a histogram of the output power for the modules shipped recently from KST. The size of glass substrates used for these modules is 910 × 910×5 mm, and the number of the total modules are around ten thousand. Except for the loss by appearance, mainly caused in the process of module constructions using these submodules, the production yields constantly exceed 95%.

As the most popular size of our products in KST was 910×455 mm, we had purchased the TCO-coated glass substrates with this size until March 2000. However, as a variation of the size was requested from several users, KST introduced a new process in these lines. In order to simplify the production

154   Yoshihisa Tawada et al.

Fig. 9.4. Fully automatic production line of the PV modules in KST

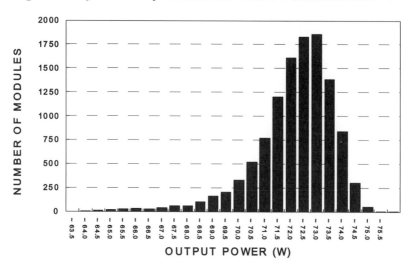

Fig. 9.5. Histogram of the large-area PV modules produced by KST

and respond to the demands of the market, KST have started to produce many
large modules with the same-size glass substrate and we cut them into several
small modules after the fabrication of the semiconductor and the back-contact
layers.

The solar modules integrated with the roofing materials were proposed and
we planned to produce them with these high-efficiency submodules. Through
the detailed design, the actual installation and the performance evaluation,
we confirmed the suitability of the proposed modules as roofing materials. It

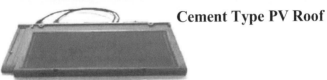

Cement Type PV Roof

Fig. 9.6. Private house with a-Si PV roof-tile modules

is said that the thin-film solar cells, particularly amorphous solar cell, can be used for the PV systems on the roofs of private houses. The most profitable situation is that any size of modules with high voltage can be prepared by cutting from the large-size modules as discussed before. Figure 9.6 shows one of the applications of a-Si solar modules in which solar submodules with 45.5 × 30 cm are put onto typical Japanese roof tiles. The PV houses roofed with these modules are, as can be seen in the figure, high-grade houses.

9.3 Performances and Production of a-Si/Thin-Film c-Si Hybrid Solar Module

At the same time, we have been developing the low-temperature deposition technologies of the thin-film poly-crystalline Si since 1993. In 1999, we achieved 9.8% efficiency of the thin film poly-Si solar cells with 3 μm thickness [7]. This solar cell keeps about 9% efficiency with decreasing the thickness from 3.5 to 1.5 μm. One of the reasons of the high performances is the natural surface texture with a "STAR Structure", which shows a dendrite-like morphology with the surface roughness on the order of 0.1 μm to few μm. High efficiency with a few μm thick is obtained by the optical confinement effect of the STAR structure. Taking advantage of the thinner layer of our thin-film poly-Si, we proposed the new hybrid cell structure of a-Si combined with thin-film c-Si. The stable efficiency of this hybrid cell with 1 cm^2 area was 12%. This hybrid module can be integrated on the substrate by the laser-

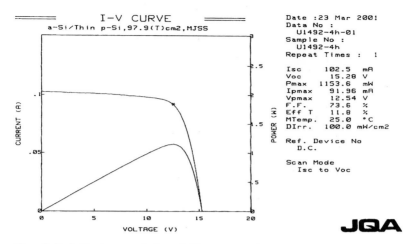

Fig. 9.7. I–V characteristics of 10×10 cm hybrid cell confirmed by JQA. The conversion efficiency, short-circuit current, open-circuit voltage, fill factor are 11.8%, 102.5 mA, 15.3 V, and 73.6%, respectively

scribe method, as well as a-Si module. We fabricated 5×5 cm module with stable 11.3% efficiency [8]. The performance of this hybrid module with an area of 10×10 cm was confirmed to be 11.8% as the initial performance by JQA, as shown in Fig. 9.7.

There existed a technology barrier to enlarge the glow-discharge deposition area in the preparation of the thin-film poly-Si. We broke through this barrier to fabricate poly-Si film with 1 m^2 area as well as a-Si.

We have developed a-Si/poly-Si hybrid (stacked) monolithic solar modules sized 910×455 mm, like the a-Si module [9]. Its cross-sectional structure of series interconnection is shown in Fig. 9.8. An initial aperture efficiency of 11.7% or an output power of 44.9 W has been achieved for 910×455 mm^2 a-Si:H/poly-Si integrated solar cell module, as shown in Fig. 9.9. The measurement of this module was performed by the dual light source solar simulator.

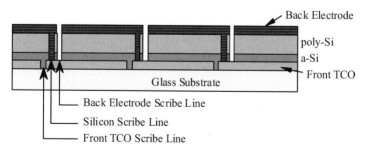

Fig. 9.8. Cross-sectional structure of series interconnection of a-Si/poly-Si hybrid (stacked) solar-cell module

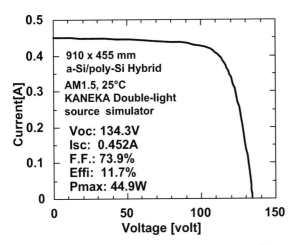

Fig. 9.9. I–V characteristics of 910×455 mm^2 a-Si/poly-Si series interconnected solar-cell module. The measurement of this module was performed by the double light source solar cell simulator (AM1.5 spectrum, 1000 W m^{-2}.) The measured aperture area was 3827 cm^2 for this module

As mentioned in the following discussion, the outdoor measurements of our module, early September in Shiga, Japan, showed good agreement with that of our indoor measurements, which were performed by a dual light source solar simulator.

The results of our first run of 266 a-Si/poly-Si hybrid modules in our pilot plant exhibit the initial average module efficiency of 11.2%, which is shown in Fig. 9.10. It was confirmed that these modules are applicable for mass production. The outdoor performances of a-Si/poly-Si hybrid (stacked) modules were investigated. For the thin-film stacked cell, it is important to measure the real module efficiency outside, since the currents of the cell are strongly affected by the shape of the spectrum. In another words, indoor measurements by a standard single light source solar simulator should be carefully performed to be fully consistent with the real outside spectrum (AM1.5). The resulting performances for four modules were listed in Table 9.1 together with a solar spectrum of our measured site (Shiga, Japan, early September, shown in Fig. 9.11). Superior coincidence between outdoor and indoor measurements, which were performed by our dual light source solar simulator, was observed as shown in Table 9.1.

As the photocurrent of a hybrid module is limited by that of a pin a-Si layer, and the spectrum in summer has a high level of the visible wavelength region, the hybrid module shows a higher output power in summer than the standard condition.

Although the hybrid-type module changes its initial performance under the sunlight as well as the a-Si module, it might be somewhat smaller than that of a-Si. Our hybrid module has also been exposed in JQA Hamanako

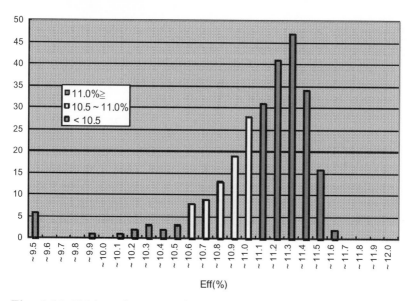

Fig. 9.10. Trial production performance distribution (initial) of 910 × 455 mm² a-Si/poly-Si hybrid modules at our pilot plant

Laboratories. Figure 9.12 shows the performance changes since May 1st, 2000. As can be seen in this figure, the light-induced change is only 8% after 10 months of outdoor exposure.

Table 9.1. The outdoor performances of four 910 × 455 mm² a-Si/poly-Si hybrid modules measured at Otsu, Shiga, Japan, on the afternoon of 5th September, 2000, together with the indoor measurements. The measured module illuminated aperture area was 3738 cm². The calibration (V_{oc}, I_{sc}) of four modules was performed to refer the measurements to standard 1000 W cm⁻², 25°C condition, which is shown in the bottom line of each module

Sample lot.	Voc (V)	Isc (A)	FF	Pmax (W)	Eff. (%)	Temp. (°C)	Sun-light intensity (sun)	Measurement
	63.77	0.871	0.755	41.96	11.22	24.6		Double-light source simulator
1	61.99	0.828	0.761	39.02		34.4	0.955	Outdoor raw data
	63.98	0.867	0.761	42.17	11.28	25	1	Outdoor (after calibration)
	63.86	0.873	0.757	42.17	11.28	24.8		Double-light source simulator
2	62.55	0.817	0.766	39.18		34.4	0.955	Outdoor raw data
	64.55	0.856	0.766	42.34	11.33	25	1	Outdoor (after calibration)
	63.30	0.896	0.745	42.25	11.30	24.4		Double-light source simulator
3	61.64	0.838	0.760	39.24		34.6	0.959	Outdoor raw data
	63.66	0.874	0.760	42.25	11.30	25	1	Outdoor (after calibration)
	63.73	0.902	0.742	42.62	11.40	24.4		Double-light source simulator
4	62.15	0.840	0.760	39.69		34.2	0.972	Outdoor raw data
	64.09	0.864	0.760	42.10	11.26	25	1	Outdoor (after calibration)

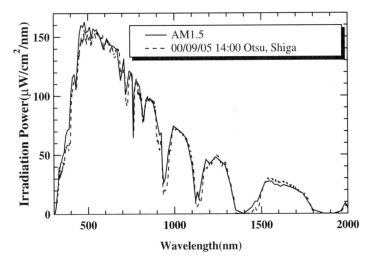

Fig. 9.11. Measured onsite solar spectrum of the afternoon of 5th of September, 2000 was shown together with the AM1.5 global spectrum

Integrating the hybrid cell by the laser-scribe technology, we fabricated the 91 × 91 cm modules showing a stable 10% efficiency.

We can easily change the production line of a-Si modules to the hybrid modules by adding the glow-discharge deposition equipment for thin-film poly-Si. As the hybrid module can be integrated by laser scribing, as well as a-Si, the equipment for a-Si modules can be utilized perfectly.

Through the marketing of the PV systems in the roof-tile applications, at least 10% module efficiency is needed to site the 3-kW modules on the south-side roof. Therefore, we invested in a new CVD system for thin c-Si in the KST-plant to introduce the hybrid technology, and began to produce

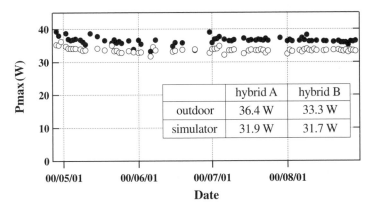

Fig. 9.12. Outdoor exposure test of hybrid module in Hamanako, JQA

Fig. 9.13. Private houses with PV roofing tiles in Japan

high-efficiency hybrid modules in April 2001. As the module efficiency was improved from 8% to 10% by changing from a-Si to the hybrid, the production capacity has been increased to 1.25 times larger and the production cost has been decreased by more than 10%.

9.4 Future Business Plan

The advantages of the thin-film PV modules are the design flexibility and the cost potential. As the modules are prepared to be cut from the maximum module with 980×950 mm, any size of modules can be applied to the roofing tiles. 910×455 mm, 910×227 mm, and 455×227 mm are the typical sizes. Figures 9.13 shows some private houses with these PV roofing tiles. In these cases the PV roofing tiles act as ordinary roofing tiles, so that the construction cost for PV system can be saved. We expect the production cost of the thin-film solar cells will be reduced as the production scale increases as in chemical plants.

We estimated the production cost of a-Si solar modules and a-Si/thin c-Si hybrid solar modules. Figure 9.14 shows the production cost reduction in accordance with the inverse square root of the production scale. The production cost might be less than Yen 200/W with the scale of more than 40 MW/y.

In order to reduce the production cost in our production line to less than half of the cost of polycrystalline-Si module our production capacity must be increased to be more than 40 MW/y. At the end of FY2002 the Japanese government will stop paying subsidies for the national residential PV rooftop dissemination program. By increasing the production volume to 100 MW/y and using the hybrid technologies with more than 12% efficiency, thin-film PV will be a major business in Japan.

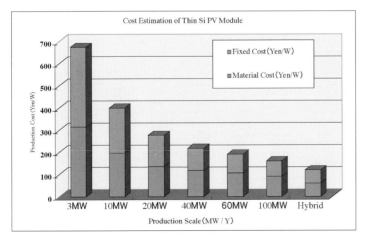

Fig. 9.14. Cost estimation of thin Si solar module as a function of production scale

Acknowledgment

This work was partially supported by the New Energy and Industrial Technology Development Organization as part of the New Sunshine Program under the Ministry of International Trade and Industry.

References

1. Kuwano, K., Tsuda, S., Onishi, M., Nishikawa, H., Nakano, S., and Imai, T. (1980): *Jpn. J. Apply. Phys.*, 20, 213.
2. Guha, S., Yang, J., Banerjee, A., Hoffman, K., and Call, J. (1999): AIP Conf. Proc. 462, 88–93.
3. Ichikawa, Y., Yoshida, T., Hama, T., Sakai, H., and Harashima, K. (1999): Technical Digest of 11th International PVSEC 49.
4. Kondo, M., Hayashi, K., Nishio, H., Kurata, S., Ishikawa, A., Takenaka, A., Nishimura, K., Nakajima, A., Yamagishi, H., and Tawada, Y. (1995): Proc.13th EU Photovoltaic Solar Energy Conf. 311.
5. Japan Quality Assurance Organization and Photovoltaic Annual Report of Research in 1998 contracted by NEDO, pp.48.
6. Tawada, Y., and Yamagishi, H. (1999): Technical Digest of 11th International PVSEC 53.
7. Yamamoto, K., Nakajima, A., Suzuki, T., Yoshimi, M., Nishio, H., and Izumina, M. (1994): *Jpn. J. Appl. Phys.*, 33, L1751.
8. Yamamoto, K., Yoshimi, M., Tawada, Y., Okamoto, Y., and Nakajima, A. (2001): *Solar Energy Mater. Sol. Cells*, 66, 117.
9. Yamamoto, K., Yoshimi, M., Suzuki, T., Nakata, T., Sawada, T., and Nakajima, A. (2000): Pro. IEEE 28th Photovoltaic Specialists Conf., p.1428.

Properties of Chalcopyrite-Based Materials and Film Deposition for Thin-Film Solar Cells

Hans-Werner Schock

The development of chalcopyrite-based solar cells started in the early 1970s, when Wagner et al. [1] realized a 12% efficient solar cell based on a $CuInSe_2$ single crystal. A few years later, Kazmerski et al. [2] were able to demonstrate the first thin-film solar cell by evaporation of $CuInSe_2$ as a compound. Even though these results were very promising it took two decades until the first commercial CIGS modules were produced. The full potential of the material was only reached after some important modifications of the growth and composition of the absorber film. The addition of Ga stabilized p-type conductivity and increased the bandgap. Sodium-containing glass, or the deliberate addition of Na to the growth process, was the key for further controlling film growth and free carrier concentration. These developments were mostly empirical and sometimes serendipitous. However, due to the involvement of more research teams world-wide basic understanding has been improved as a prerequisite for reproducible manufacturing processes.

10.1 Cu-Chalcopyrite Compounds

The Cu-chalcopyrite compound $CuInSe_2$ and its alloys $Cu(In,Ga)(Se,S)_2$ provide the absorber material for the, to date, most efficient thin-film solar cells. A high-power conversion efficiency of close to 19% obtained with an alloy $Cu(In,Ga)Se_2$ (with a Ga/(Ga+In)-ratio $\simeq 0.2$) reaches the level obtained with conventional wafer-based, polycrystalline-silicon solar-cell technology [3]. The $Cu(In_{1-x}Ga_x)(Se_{1-y}S_y)_2$ alloy system provides vast design options for processing and properties for the development of high-efficiency solar cells with high open-circuit voltage solar cells as the bandgap energy E_g of $CuInSe_2$ increases upon alloying with Ga and/or S [4]. Besides the technologically advantageous features including the outstanding radiation hardness of $CuInSe_2$ and its alloys [5, 6], the Cu-chalcopyrites also attract considerable scientific interest because of their unusual defect physics: e.g., the ability to form electronically inactive defect complexes explains the great tolerance of these materials to deviations from stoichiometry and to foreign impurities [7].

10.1.1 Material Properties

$CuInSe_2$ and $CuGaSe_2$, the compounds that form the alloy $Cu(In,Ga)Se_2$, belong to the semiconducting I–III–VI$_2$ compounds. Their lattice elements are tetrahedrally coordinated like in all adamantine (diamond-like) semiconductors. The tetragonal chalcopyrite structure of, for example, $CuInSe_2$ is derived from the cubic zinc blende structure of II–VI materials like ZnSe by occupying the Zn sites alternately with Cu and In atoms. In the chalcopyrite structure each I (Cu) or III (In) atom has four bonds to the VI atom (Se). In turn each Se atom has two bonds to Cu and two to In. Because the strengths of the I–VI and III–VI bonds are different, the ratio of the lattice constants c/a is not exactly 2, i.e., it varies from 2.01 in $CuInSe_2$, to 1.96 in $CuGaSe_2$.

The absorption coefficients of chalcopyrite compounds in comparison with other semiconductors are depicted in Fig. 10.1. The specifically high absorption coefficient makes the chalcopyrite compounds well suited for thin-film solar cells.

Further important properties for the operation of solar cells are charge-carrier mobilities, diffusion length, and minority-carrier lifetimes. These quantities depend very much on the preparation conditions of the thin films. Electron mobilities in single crystals reach $1000 \ cm^2V^{-1}s^{-1}$ [8, 9]. In connection with lifetimes of several ns, the electron diffusion length is on the order of a few μm. With an absorption length below one micrometer (Fig. 10.1), CIGS-based solar cells could reach very high efficiencies.

A further important feature of the material is the role of the grain boundaries. Up to efficiencies in the 15% range, grain size does not seem to play an important role for device performance. However, a profound understanding of grain-boundary properties is still lacking.

10.1.2 Phase Diagram

Due to the multitude of elements and compounds involved in the formation of $Cu(In_{1-x}Ga_x)(Se_{1-y}S_y)_2$ defects and growth of films are extremely complex. The phase diagrams have been extensively investigated by Gödecke et al. [10]. These investigations had a special focus on temperatures and compositions relevant for the preparation of thin films.

Figure 10.2 displays a simplified version of the ternary phase diagram, which comprises all ternary Cu-In-Se compounds. This ternary phase diagram can be reduced to a simpler pseudobinary phase diagram along the tie line between Cu_2Se and In_2Se_3 (dotted line in Fig. 10.2). The bold points along this line represent the range of photovoltaic-quality material. By combining these two compounds, $CuInSe_2$ can be formed:

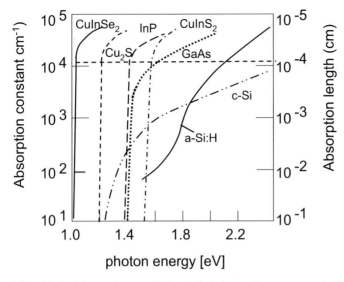

Fig. 10.1. Absorption coefficient of chalcopyrite compounds together with data of other semiconductors applied in photovoltaics

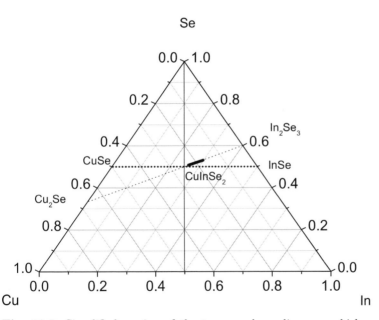

Fig. 10.2. Simplified version of the ternary phase diagram, which comprises all ternary Cu-In-Se compounds. The bold points represent the range of photovoltaic-quality material

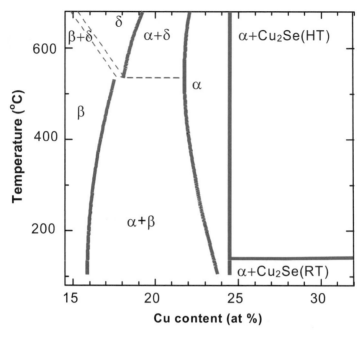

Fig. 10.3. Pseudobinary cut Cu_2Se-In_2Se_3 of the ternary phase diagram in Fig. 10.2. The α-phase has a narrow range of existence

$$x(Cu_2Se) + (1 - x)In_2S_3) \ , \qquad 0 < x < 1 \ .$$

Figure 10.3 shows the relevant portion of the phase diagram of $CuInSe_2$ given by the phase diagram in Fig. 10.2 [16]. There are four different phases which have been found to be relevant in this range: the α-phase ($CuInSe_2$), the β-phase ($CuIn_3Se_5$), the δ-phase (the high-temperature sphalerite phase) and $Cu_{2-y}Se$. An interesting point is that the phases adjacent to the α-phase have a similar structure. The β-phase is actually a defect chalcopyrite phase built by ordered arrays of defect pairs (Cu vacancies V_{Cu} and In-Cu antisites In_{Cu}). Similarly, $Cu_{2-y}Se$ can be viewed as being constructed from the chalcopyrite by using Cu-In antisites Cu_{In} and Cu interstitials Cu_i. The transition to the sphalerite phase arises from disordering the cation (Cu, In) sub-lattice, and leads back to the zinc blende structure.

The existence range of the α-phase in pure $CuInSe_2$ Cu_2Se-In_2Se_3 at room temperature extends from a Cu content of 24% to 24.5% on the quasibinary tie line [16]. Thus, the existence range of single-phase $CuInSe_2$ is astonishingly small and does not even include the stoichiometric composition of 25% Cu. The Cu content of absorbers for efficient thin-film solar cells typically varies between 22 and 24 at % Cu. At the growth temperature of 500–550°C this region lies within the single-phase region of the α-phase. However, at room temperature it lies in the two-phase $\alpha + \beta$ region of the equilibrium phase diagram in [16]. Hence, there is a tendency for phase separation in $CuInSe_2$ after deposition. Fortunately, it turns out that partial replacement of In with Ga, as well as the use of Na-containing substrates, considerably widens the single-phase region in terms of $(In + Ga)/(In + Ga + Cu)$ ratios [18]. Thus, the studies of the phase diagram, especially in connection with the addition of Ga and Na explain the substantial improvements actually achieved in recent years by adding Na to the growth process either from Na-containing substrates or from Na-containing precursor layers, as well as by the use of $Cu(In,Ga)Se_2$ alloys.

10.1.3 Defects and Impurities

The role of defects in the ternary compound $CuInSe_2$, and even more so in $Cu(In,Ga)Se_2$, is of special importance because of the large number of possible intrinsic defects [11] and the role of deep recombination centers in the performance of the solar cells [12]. The challenge of defect physics in $Cu(In,Ga)Se_2$ is to explain three unusual effects in this material:

- doping of $Cu(In,Ga)Se_2$ with native defects;
- the electrical tolerance to large off-stoichiometries;
- the electrically neutral nature of the structural defects.

It is obvious that the explanation of these effects significantly contributes to the explanation of the photovoltaic performance of this material. It is known that the doping of $CuInSe_2$ is controlled by intrinsic defects. Samples with p-type conductivity are grown if the material is Cu-poor and annealed under high Se vapor pressure, whereas Cu-rich material with Se deficiency tends to be n-type [13, 14]. Thus, the Se vacancy V_{Se} is considered to be the dominant donor in n-type material (and also the compensating donor in p-type material), and the Cu vacancy V_{Cu} the dominant acceptor in Cu-poor p-type material.

In contrast to the binary compounds, where small deviations from stoichiometry cause drastic changes of the electronic properties, the ternary compounds, in particular $CuInSe_2$, are much more tolerant. Note that the Cu content of device-quality $CuInSe_2$ or $Cu(In,Ga)Se_2$ absorbers varies typically between 22 and 24 at % Cu. Thus, these films are markedly Cu-poor but nevertheless maintain excellent semiconducting properties. In terms of point defects, a nonstoichiometry of 1% would correspond to a defect concentration of roughly 10^{21} cm^{-3}. This is by about five orders of magnitude more than the acceptable density of recombination centers in a photovoltaic absorber material and on the order of the net doping concentration of about 10^{17} cm^{-3} that is useful for the photovoltaic active part of a solar cell. Even if we allow a degree of compensation of 99%, the respective densities of donors and acceptors would be only in the 10^{19} cm^{-3} range. The virtual number of defects calculated from off-stoichiometry has to be brought down to reasonable quantities that are compatible with the good electronic quality that is required for a photovoltaic device. Deviations of the valence stoichiometry, i.e., the proper Se/metals ratio according to the pseudobinary tie line in Fig. 10.3, however, cause significant changes in the electronic properties. If the material is grown under Se saturation, the Se content in the film is self-adjusting, i.e., excess Se reevaporates.

There are two ways that could lead to a material with good electronic properties at high stoichiometry deviations: Either the nonstoichiometry is accommodated in a secondary phase that is not harmful to the photovoltaic performance, or the off-stoichiometry-related defects are electronically inactive. In principle, $CuInSe_2$ could realize both possibilities. The two point defects that are related to an overall Cu-poor composition are the Cu-vacancy V_{Cu} (a shallow acceptor) and the In-Cu antisite In_{Cu} (a deep double donor). According to first-principle calculations these two defects form a defect complex $(2V_{Cu}, In_{Cu})$ that is electronically neutral and has no energy levels within the energy gap of $CuInSe_2$ [7]. Therefore, these complexes do not contribute to the electronically active defects. Ordered arrays of this complex can be considered as the building blocks of a series of Cu-In-Se compounds like $CuIn_3Se_5$, $CuIn_5Se_8$ [9, 15], i.e., the secondary phases that delimit the single-phase region of $CuInSe_2$ (α-phase) towards the Cu-poor side [16]. Due to the lower Cu concentration these compounds have a lower valence-band energy and hence a wider bandgap. Segregations of these phases are found in Cu-poor single crystals [17].

At room temperature, the equilibrium phase diagram of pure $Cu_xIn_{1-x}Se_2$ [10] predicts a two-phase $\alpha + \beta$ region for Cu contents below 24%, i.e., the stability region of the α-phase is quite narrow. The situation with respect to secondary phase segregations is a little different in thin films compared to pure single-crystal $CuInSe_2$. First, in these films about 20–30% of the In is replaced by Ga and, second, these films grow on Na-containing glass and, consequently, contain around 0.1 at% Na that diffuses out of the glass during deposition. Both, the incorporation of Ga and Na inhibit the ordering of the defect complexes and hence the formation of secondary phases in the bulk of the thin film [18].

The surface of In-rich films exhibits the 1-3-5 composition of the $Cu(In,Ga)_3Se_5$ defect phase [19], i.e., the surface is always more Cu-poor than the bulk of the material. Most likely, the Cu-poor surface of these thin films is not the consequence of the segregation of a secondary bulk phase but rather results from the accumulation of Cu-deficiency-related defects and defect complexes [18]. The precise properties of this surface defect layer are not yet known. However, it is of high importance that the bandgap at the surface of the $Cu(In,Ga)Se_2$ thin films is somewhat larger than the bandgap in the bulk of the material [19].

On the Cu-rich side of the single-phase region of the phase diagram, Cu_2Se segregates as a secondary phase. In thin films, Cu_2Se can be easily removed by dissolving in potassium cyanide (KCN) solution leaving behind a $CuInSe_2$ or $Cu(In,Ga)Se_2$ film with the proper stoichiometry. It is one of the puzzles of this material that films prepared in such a way exhibit good electronic quality (judged, e.g., from photoluminescence experiments [20, 21]). However, solar cells made from material that is grown in such a way are far less efficient than those with a slightly Cu-deficient absorber.

The related compound $CuInS_2$ recently gained attention as an absorber material for thin-film solar cells [22]. Growth processes in this case rely on the Cu-rich growth of the films in connection with subsequently removing excess CuS by dissolving in KCN solution.

10.1.3.1 Models for Defect Structures

By calculating the metal-related defects in $CuInSe_2$ and $CuGaSe_2$, the authors of [7] found that the defect-formation energies for some intrinsic defects are low and depend on the chemical potential of the components (i.e., on the composition of the material) as well as on the electrochemical potential of the electrons (i.e., the Fermi level). For V_{Cu} in Cu-poor and stoichiometric material, a negative formation energy is calculated. This would imply the spontaneous formation of large numbers of those defects if the Fermi level changes. Low (but positive) formation energies are also found for the Cu-on-In antisite Cu_{In} in Cu-rich material (this defect is a shallow acceptor that could be responsible for the p-type conductivity of Cu-rich, non-Se-deficient $CuInSe_2$). The dependence of the defect-formation energies on the electron Fermi level could explain the strong tendency of $CuInSe_2$ to self-compensation and the difficulties of achieving extrinsic doping. The work of [7] provides a theoretical basis for the calculation of defect-formation energies and defect-transition energies, which exhibit good agreement with experimentally obtained data.

Further important results are the formation energies of some defect complexes such as $(2V_{Cu}, In_{Cu})$, (Cu_{In}, In_{Cu}), and $(2Cu_i, Cu_{In})$, where Cu_i is an interstitial Cu atom [23]. These formation energies are even lower than those of the corresponding isolated defects. Interestingly, $(2V_{Cu}, In_{Cu})$ does not exhibit an electronic transition within the forbidden gap, in contrast to the isolated In_{Cu} antisite, which is a deep recombination center. As the $(2V_{Cu}, In_{Cu})$ complex is most likely to occur in In-rich material, it can accommodate a large amount of excess In (or likewise Cu deficiency) and, at same time, does not significantly influence the electrical performance of the material. Furthermore, ordered arrays of this complex can be thought as of the building blocks of the In-rich Cu-In-Se compounds such as $CuIn_3Se_5$ and $CuIn_5Se_8$ [23].

10.1.3.2 The Role of Na

The influence of Na on the growth of $Cu(In,Ga)Se_2$ films was found some years ago [24]–[26]. The early thin-film devices were fabricated on glass or Na-free glass substrates). When soda lime glass is used, Na from the glass substrate diffuses into the absorber. In order to better control the sodium content it is also deliberately incorporated in the film by the use of Na-containing precursors such as Na_2Se [27, 28], Na_2O_2 [26], and NaF [29]. Among different alkali precursors Na-containing precursors yielded the best cell efficiencies. The most obvious effects of Na incorporation are better film morphology and higher conductivity of the films [26]. Furthermore, the incorporation of Na reduces the defect concentration of the absorber films [30, 31].

The explanations for the effect of Na are manifold, and it is most likely that the interaction of Na with the growing film has a variety of consequences. A basic process appears to be the reaction of Na with Se and the formation of intermediate compounds. During film growth, the incorporation of Na leads to the formation of $NaSe_x$ compounds. This slows down the growth of $CuInSe_2$ and could at the same time facilitate the incorporation of Se into the film [32, 33]. A similar, however, less-pronounced effect can be ascribed to the presence of Ga [34]. Also, the widening of the existence range of the α-$(CuInSe_2)$ phase in the phase diagram, discussed above, as well as the reported larger tolerance to the $Cu/(In + Ga)$ ratio of Na-containing thin films, could be explained in this picture. Furthermore, the higher conductivity of Na-containing films could result from the diminished number of compensating V_{Se} donors. The presence of Na has a significant influence on the kinetics during formation of $CuInSe_2$ films from stacked elemental layers. Thin-film calorimetry [35] revealed that Na inhibits the growth of $CuInSe_2$ at temperatures below 380°C. The retarded phase formation due to stabilization of the intermediate binary Cu-compounds is responsible for the better morphology in the case of Na-containing samples.

The above explanations deal with the role of Na during growth. However, the amount of Na found in device-quality $Cu(In,Ga)Se_2$ films is of the order 0.1 at %, which is a concentration of 10^{20} cm^{-3} [36]. This amount is far beyond the free carrier concentration. The change of effective doping resulting from Na incorporation, is achieved at concentrations of $\sim 10^{16}$ cm^{-3}, four orders of magnitude below the absolute Na content. It has long been believed that the majority of the Na is situated at the film surface and the grain boundaries. Confirmation of this hypothesis was recently found by [37] with the help of high-resolution scanning Auger electron spectroscopy.

A further explanation, put forward in [38], is that Na promotes oxygenation and passivation of grain boundaries. This could account for the observed enhancement of the net film doping by Na incorporation, through the diminished positive charge at the grain boundaries. It has, in fact, been observed that oxygenation of the surfaces of Na-containing films is much stronger compared to Na-free films [39].

Another interpretation of the beneficial effect of Na is based on the incorporation of Na into the $Cu(In,Ga)Se_2$ lattice [36]. Niles and co-workers [36] identified Na-Se bonds by means of XPS and concluded that the Na is built into the lattice, replacing In or Ga. The extrinsic defect $Na_{In/Ga}$ should then act as an acceptor to improve the p-type conductivity. The incorporation of Na into the $Cu(In,Ga)Se_2$ lattice is supported by X-ray diffraction measurements that indicate an increased volume of the unit cell. In this case Na in a

Cu site could prevent the formation of the deep double donor InCu. Na driven into epitaxial Cu(In,Ga)Se$_2$ films at a temperature of 550°C decreases the degree of compensation by up to a factor 104 [40]. These finding are attributed to a Na-enhanced reorganization of the defects, which allows them to build electrically passive clusters.

Despite the importance of Na for CIGS film formation and properties, the benefit is far from being explained in terms of simple models. In view of the amounts of Na (~ 0.1 at %) necessary for optimum film preparation, arguments based on its effect on film growth and control of defect formation are more conclusive than those based on the incorporation of Na into the completed film.

10.2 Alloys

The present design of a standard cell is the result of an historic development based upon empirical optimization. Improved knowledge about the material and devices now allow systematic developments. The exploitation of the chalcopyrite compound material system expanded considerably in the recent past. Alloys of multinary chalcopyrite compounds allow all kinds of bandgap variations in the absorber film. Figure 10.4 presents an overview about the possibility of bandgap engineering and the design of wide-bandgap devices in the Cu(In,Ga,Al)(S,Se)$_2$ system. The bandgap energies of I-III-VI$_2$ chalcopyrites are considerably smaller than those of their binary analogs (this is the binary material where the I/III elements are replaced by their average II element; thus ZnSe is the binary analog of CuGaSe$_2$ and Zn$_{0.5}$Cd$_{0.5}$Se that of CuInSe$_2$). This difference is because the Cu 3d band, together with the Se 4p band, forms the uppermost valence band in the Cu-chalcopyrites, which is not so in II-VI compounds. However, the system of copper chalcopyrites displayed in Fig. 10.4 covers the visible spectrum with a wide range of band-gap energies from 1.04 eV in CuInSe$_2$ up to 2.7 eV in CuAlSe$_2$. Figure 10.4 summarizes lattice constants and bandgap energies of this system. Any desired alloys between these compounds can be produced, as there is no miscibility gap in the entire system.

The different compounds have different electronic properties. In particular, the situation of the phase diagram of all chalcopyrite compounds is not necessarily as favorable as for CuInSe$_2$ [41, 42].

10.2.1 Growth Methods for Thin Films

Even though the first photovoltaic device based on CuInSe$_2$ was a heterojunction with evaporated CdS on a CIS single crystal [1], high-efficiency single-crystal devices have never been reproduced. Various attempts to produce epitaxial films as a reference contribute to the basic understanding of the material [43, 44] but did not result in the development of very high efficiency devices.

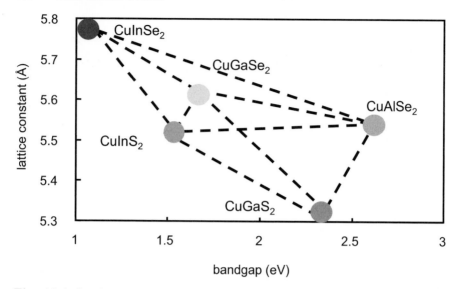

Fig. 10.4. Lattice constants versus bandgap of ternary chalcopyrite compounds that could form multinary alloys. Complete miscibility allows design of wide-gap and graded-gap devices

Photovoltaic-grade $Cu(In,Ga)Se_2$ films have a slightly In-rich overall composition. The allowed stoichiometry deviations are astonishingly large, yielding a wide process window with respect to composition. This great tolerance of CIS to deviations in the stoichiometry is a prerequisite for the success of CIGS solar-cell production. Devices with efficiencies above 14% are obtained from absorbers with $(In + Ga)/(In + Ga + Cu)$ ratios between 52 and 64%, if the sample contains Na. Cu-rich $Cu(In,Ga)Se_2$ shows the segregation of a secondary $Cu_{2-y}Se$ phase preferentially at the surface of the absorber film. The metallic nature of this phase does not allow the formation of efficient heterojunctions. Even after removal of the secondary phase from the surface by dissolving in KCN, the utility of Cu-rich material for photovoltaic applications is limited, probably due to the high doping density of 10^{18} cm^{-3} in the bulk and the surface defects. However, the importance of the Cu-rich composition is given by its role during film growth. Cu-rich films have grain sizes in excess of 1 μm whereas In-rich films have much smaller grains. A model for the film growth under Cu-rich compositions comprises the role of $Cu_{2-y}Se$ as a flux agent during the growth process of coevaporated films [45]. This model for the growth of $Cu(In,Ga)Se_2$ in the presence of a quasiliquid surface film of Cu_ySe is highlighted in Fig. 10.5. In the presence of Cu_xSe, crystal growth is determined by this quasiliquid surface phase that helps to incorporate In and Se in a liquid-solid growth mechanism.

For $Cu(In,Ga)Se_2$ prepared by selenization, the role of $Cu_{2-y}Se$ is similar [46]. Cu_xSe mediates the growth of the crystallites, during the selenization

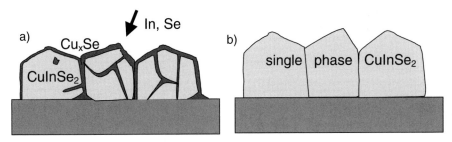

Fig. 10.5. Growth model in the presence of Cu_xSe: a) liquid-solid growth due the presence of Cu_2Se, b) fully crystallized slightly In-rich film

from solid Se films (RTP) or elemental evaporation from the gas phase or therefore growth processes for high quality have to go through a copper-rich stage and end In-rich.

10.2.2 Vacuum Evaporation Methods

The first thin-film cells were produced by evaporation of $CuInSe_2$ as a compound together with elemental selenium [2, 47]. Complete evaporation of a well-defined amount of the metals together with the coevaporation of Se leads to the correct composition of the film and a device-quality film without additional process control [48]. For scalable production processes and for a good control of film growth other methods have to be applied.

10.2.2.1 Coevaporation Processes

The absorber material yielding the highest efficiencies is $Cu(In,Ga)Se_2$ with a Ga/(Ga + In) ratio of $\sim 20\%$, prepared by coevaporation from elemental sources. Figure 10.6 sketches a coevaporation setup as used for the preparation of laboratory-scale solar cells and minimodules.

The process requires a substrate temperature of $\sim 550°C$ for a certain time during film growth, preferably towards the end of growth. One advantage of the evaporation route is that material deposition and film formation are performed during one processing step. A high reproducibility is obtained if the evaporation rates are well controlled. A feedback control based on electron impact emission spectroscopy (EEIS), a quadrupole mass spectrometer (QMA) or an atomic absorption spectrometer (AAS) controls the rate of each source. The composition of the deposited material with regard to the metals corresponds to their evaporation rates, whereas Se is always evaporated in excess. An additional possibility to control the composition is to monitor the emissivity of the films [49]. At the transition between In-rich and Cu-rich compositions the thermal emission changes, resulting in a change of substrate temperature at constant heating power or a change of heating power while

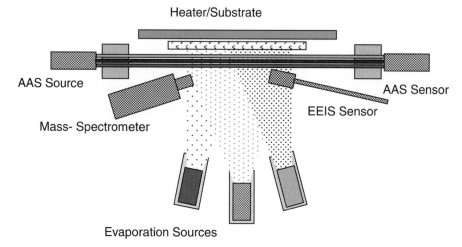

Fig. 10.6. Coevaporation setup as used for the preparation of laboratory-scale solar cells and minimodules including various methods for process control

keeping a constant substrate temperature. This behavior allows us to adjust the process precisely.

This precise control over the deposition rates allows for a wide range of variations and optimizations with different substeps or stages for film deposition and growth. In the simplest single-step process, all rates as well as the substrate temperature are kept constant during the whole process.

Advanced preparation sequences always include a Cu-rich stage during the growth process and end with an In-rich overall composition in order to combine the large grains of the Cu-rich stage with the otherwise more favorable electronic properties of the In-rich composition. Figure 10.7 illustrates two possibilities, in Fig. 10.7a the so-called Boeing or *bilayer process* [50, 51], which starts with the deposition of Cu-rich $Cu(In,Ga)Se_2$ and ends with an excess In rate. Another possibility is the *inverted process*, where first $(In,Ga)_2Se_3$ (likewise In, Ga, and Se from elemental sources to form that compound) is deposited at a lower temperatures (typically around 300°C). Then Cu and Se are evaporated at an elevated temperature until an overall composition close to stoichiometry is reached [52]. This process leads to a smoother film morphology than the process that starts Cu-rich. The most successful modification of the inverted process is the so-called *three-stage process* [53] shown in Fig. 10.7b. In this process In, Ga, and Se are again evaporated at the end of an inverted process to ensure the overall In-rich composition of the film, even if the material is Cu-rich during the second stage. The three-stage process currently leads to the best solar cells. In general, variations of the Ga/In ratio during deposition, allow the design of graded-bandgap structures to be accomplished [54, 55].

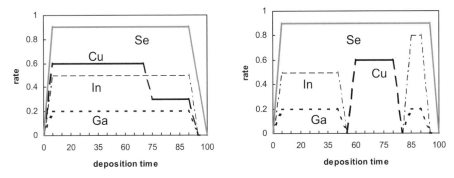

Fig. 10.7. Rate profiles of different coevaporation processes: (left) "Boeing process", (right) inverted three-stage process

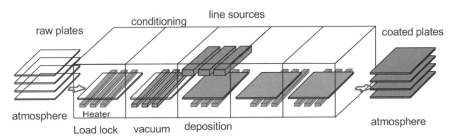

Fig. 10.8. Inline processes with line sources for large-area substrates

Coevaporation can be realized easily only for relatively small areas. Great efforts have been invested in the development of an inline coating system that allow us to continuously coat substrates with a width of 0.6 m. Figure 10.8 shows the schematic view of such a deposition system, where the substrates can be continuously processed in an air-to-air loadlock system [56].

10.2.3 Reactive Film Formation

The complexity of the equipment for coevaporation processes can be circumvented if the layer can be formed from elemental of compound precursor films. These precursor films can be put down by any method for thin-film deposition. Figure 10.9 illustrates the different possibilities for film formation by reactive processes.

10.2.3.1 Selenization from Se Vapor

The second class of absorber preparation routes is based on the separation of deposition and compound formation into two different processing steps. High efficiencies are obtained from absorbers prepared by selenization of metal precursors in H_2Se [57]–[59] and by rapid thermal processing of stacked elemental

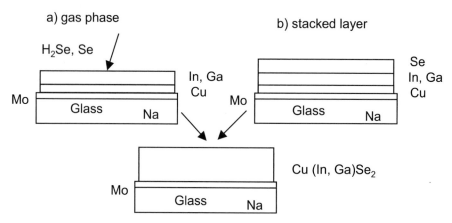

Fig. 10.9. Illustration of the sequential process. a) Stacked metal layers are selenized and converted into CuInSe$_2$ in a H$_2$Se atmosphere. b) Se is provided by a solid layer, reaction kinetics is controlled by the annealing temperature

layers in a Se atmosphere. These sequential processes have the advantage that approved large-area deposition techniques such as sputtering can be used for the deposition of the materials. The Cu(In,Ga)Se$_2$ film formation then requires a second step, the selenization.

The first large-area modules were prepared by the selenization of metal precursors in the presence of H$_2$Se more than ten years ago [60]. Today, a modification of this process is providing the first commercially available Cu(In,Ga)Se$_2$ solar cells, manufactured by Siemens Solar Industries. This process is schematically drawn in Fig. 10.9a. First, a stacked layer of Cu, In, and Ga is sputter deposited on the Mo-coated glass substrate. Then selenization takes place under H$_2$Se. To improve device performance, a second thermal process under H$_2$S is added, resulting in an absorber that contains a Cu(In,Ga)(S,Se)$_2$ surface layer.

10.2.4 Annealing of Stacked Elemental Layers

A variation of this method that avoids the use of the toxic H$_2$Se during selenization is the rapid thermal processing of stacked elemental layers shown in Figure 10.9b [46]. Here, the precursor includes a layer of evaporated elemental Se. The stack is then selenized by a rapid thermal process (RTP) in either an inert or a Se atmosphere. The highest efficiencies are obtained if the RTP is performed in an S-containing atmosphere (either elemental S or H$_2$S).

On the laboratory scale, the efficiencies of cells made by these preparation routes are smaller by about 3% (absolute) as compared with the record values. However, on the module level, coevaporated and sequentially prepared absorbers have about the same efficiency. Sequential processes need two or even three stages for absorber completion. These additional processing steps may counterbalance the advantage of easier element deposition by sputtering. Also, the detailed and sophisticated control over composition and growth achieved during coevaporation is not possible for the selenization process. The reaction kinetics is controlled by the time and temperature of the annealing process. Fortunately, the distribution of the elements within the film grown during the selenization process turns out to be close to a certain optimum, especially if the process includes the sulphurization stage. Since the formation of $CuInSe_2$ is much faster than that of $CuGaSe_2$, and because the formation of the ternary chalcopyrite phase starts at the reaction front with the Se at the surface, Ga is concentrated towards the back surface of the film. An increasing Ga content implies an increase in bandgap energy. This introduces a so-called back-surface field, improving carrier collection at the same time as minimizing back-surface recombination. In turn, S from the sulphurization step is incorporated at the front surface of the film, where it reduces recombination losses and also increases the absorber bandgap in the space-charge region of the heterojunction.

10.2.5 Epitaxy, Chemical Vapor Deposition, and Vapor Transport Processes

Besides selenization and coevaporation, other deposition methods have been studied, either to obtain films with very high quality or to reduce the cost of film deposition on large areas. Methods that are used to form epitaxial III-V compound films, such as molecular beam epitaxy (MBE) [44] or metal organic chemical vapor deposition (MOCVD) [61] have revealed interesting features for fundamental studies, such as phase segregation and defect formation, but cannot be used to form the base material for high-efficiency solar cells. Chemical vapor transport of Cu_2Se and Ga_2Se_3 was applied to successfully grow $CuGaSe_2$ thin films [62].

10.2.6 Other Techniques

Attempts to develop so-called low-cost processes include electrodeposition, [63]–[65] screen printing, and particle deposition [66, 67]. Electrodeposition can be carried out in either one or two steps. Codeposition of all elements requires adjustment of concentrations and complexing agents. The crucial step in all cases is the final film formation in a high-temperature annealing process. The recrystallization process competes with the decomposition of the material, therefore process optimization is quite difficult. Cells with good efficiencies were obtained by a hybrid process combining electrodeposition of a Cu-rich $CuInSe_2$ film and subsequent conditioning by a vacuum-evaporation step of In(Se) [68, 69]. Again, annealing of the films at elevated temperature in a suitable atmosphere is an important step to the low-cost deposition processes of the precursor film.

References

1. Wagner, S., Shay, J. L., Migliorato, P. and Kasper, H.M., 'CuInSe$_2$/CdS heterojunction photovoltaic detectors', *Appl. Phys. Lett.* 25 (1974) 434–435.
2. Kazmerski, L.L., Ayyagari, M.S., Juang, Y.J., R.P. Patterson, R.P., 'Growth and Characterisation of Thin Film Compound Semiconductor Heterojunctions', *J. Vac. Sci. Technol.* 13 (1977) 65.
3. Contreras, M., Egaas, B., Ramanathan, K., Hiltner, J., Swartzlander, A., Hasoon, F., and Noufi, R., *Prog. Photovolt. Res. Appl.* 7 (1999) 311.
4. Shay, J.L. and Wernick, J.H., *Ternary Chalcopyrite Semiconductors: Growth, Electronic Properties, and Applications* (Pergamon Press, Oxford, 1975).
5. Jasenek, A., Rau, U., Hahn, T., Hanna, G., Schmidt, M., Hartmann, M., Schock, H.E., Werner, J.H., Schattat, B., Kraft, S., Schmid, K.-H., and Bolse, W., *Appl. Phys. A* 70 (2000) 677.
6. Jasenek, A., and Rau, U., *J. Appl. Phys.* 90 (2001) 650.
7. Zhang, S.B., Wei, S.H., Zunger, A., and Katayama-Yoshida, H., *Phys. Rev. B* 57 (1998) 9642.
8. Wasim, S.M., *Solar Cells* 16 (1986).
9. Wasim, S.M., Rincon, C., Marin, G., Marquez, R., Sanchez Perez, R., Guevara, R., Delgado, J.M. Nieves, L., 'Growth, Structural Characterisation, and Optical Bandgap Anomaly in Cu-III$_3$-VI$_5$ and Cu-III$_5$-VI$_8$ Ternary compounds', *Mater. Res. Soc. Symp. Proc.* 668 (2001).
10. Gödecke, T., Haalboom, T., Ernst, F., Phase Equilibria of Cu-In-Se, *Z. Metallkd.* 91 (2000) 622–662.
11. Cahen, D. (1987), 'Some thoughts about defect chemistry in ternaries', *Proc. 7th. Int. Conf. on Ternary and Multinary Compounds, Mater. Res. Soc.*, Pittsburgh, 433–442.
12. Burgelman, M., Engelhardt, F., Guillemoles, J.-F., Herberholz, R., Igalson, M., Klenk, R., Lampert, M., Meyer, T., Nadenau, V., Niemegeers, A., Parisi, J., Rau, U., Schock, H.-W., Schmitt, M., Seifert, O., Walter, T., and Zott, S. (1997), 'Defects in Cu(In,Ga)Se$_2$ semiconductors and their role in the device performance of thin-film solar cells', *Prog. Photovoltaic Res. Appl.* 5 (1977) 121–130.
13. Migliorato, P., Shay, J. L., Kasper, H.M., and Wagner, S., 'Analysis of the electrical and luminescent properties of CuInSe$_2$', *J. Appl. Phys.* 46 (1975) 1777–1782.
14. Noufi, R., Axton, R., Herrington, C., and Deb, S.K., 'Electronic properties versus composition of thin films of CuInSe$_2$', *Appl. Phys. Lett.* 45 (1984) 668–670.
15. Zhang, S.B., Wei, S.H., and Zunger, A., *Phys. Rev. Lett.* 78 (1977) 4059.
16. Haalboom, T., Gödecke, T., Ernst, F., Rühle, R., Herberholz, R., Schock, H.-W., Beilharz, C., and Benz, K.W., *Inst. Phys. Conf. Ser.* 152E (1998) 249.
17. Hornung, M., Benz, K.W., Margulis, L., Schmid, D., and Schock, H.-W., 'Growth of Bulk Cu$_{0.85}$In$_{1.05}$Se$_2$ and Characterization on a Micro Scale', *J. Cryst. Growth* 154 (1995) 315.
18. Herberholz, R., Rau, U., Schock, H.-W., Haalboom, T., Gödecke, T., Ernst, F., Beilharz, C., Benz, K.W., and Cahen, D., *Eur. Phys. J. AP* 6 (1999) 131.
19. Schmid, D., Ruckh, M., Grunwald F., and Schock, H.-W., *J. Appl. Phys.* 73 (1993) 2902.
20. Zott, S., Leo, K., Ruckh, M., and Schock, H.-W., *J. Appl. Phys.* 82 (1977) 356.

21. Wagner, M., Dirnstorfer, I., Hofmann, D.M., Lampert, M.D., Karg, F., and Meyer, B.K., *phys. stat. sol.* (a) 167, 131 (1998); Dirnstorfer, I., Wagner, M., Hoffmann, D.M., Lampert, M.D., Karg, F., and Meyer, B.K., *phys. stat. sol.* (a) 168 (1998) 163.

22. Klaer, J., Siemer K., Luck. I., Braunig, D., '9.2% efficient $CuInS_2$ mini-module', *Thin Solid Films*, 387 (2001) 169–171.

23. Zhang, S.B., Wei, S.H., and Zunger, A., 'Stabilization of ternary via ordered arrays of defect pairs', *Phys. Rev. Lett.* 78 (1977) 4059–4062.

24. Hedström, J., Ohlsen, H., Bodegard, M., Kylner, A., Stolt, L., Hariskos, D., Ruckh, M., and Schock, H.-W., '$ZnO/CdS/Cu(In,Ga)Se_2$ thin film solar cells with improved performance', Conf. Record 23rd. IEEE Photovoltaic Specialists Conf., Louisville, IEEE Press, Piscataway (1993) 364–371.

25. Stolt, L., Hedström, J., Kessler, J., Ruckh, M., Velthaus K.O. and Schock H.W., '$ZnO/CdS/CuInSe_2$ thin-film solar cells with improved performance', *Appl. Phys. Lett.* 62 (1993) 597–599.

26. Ruckh, M., Schmid, D., M. Kaiser, R. Schäffler, Walter, T., and Schock, H.-W., 'Influence of Substrates on the Electrical Properties of $Cu(In,Ga)Se_2$ Thin Films', *Solar Energy Mater. Solar Cells* 41/42 (1996) 335.

27. Holz, J., Karg, F., and v. Phillipsborn H. (1994), 'The effect of substrate impurities on the electronic conductivity in CIGS thin films, Proc. 12th. European Photovoltaic Solar Energy Conf., Amsterdam, H.S. Stephens & Associates, Bedford, 1592–1595.

28. Nakada, T., Iga, D., Ohbo, H., and Kunioka, A., 'Effects of sodium on $Cu(In,Ga)Se_2$-based thin films and solar cells', *Jpn. J. Appl. Phys.* 36 (1997) 732–737.

29. Contreras, M.A., Egaas, B., Dippo, P., Webb, J., Granata, J., Ramanathan, K., Asher, S., Swartzlander, A., and Noufi, R. (1997a), 'On the role of Na and modifications to $Cu(In,Ga)Se$ absorber materials using thin MF (M = Na, K, Cs) precursor layers', Conf. Record. 26th. IEEE Photovoltaic Specialists Conf., Anaheim, IEEE Press, Piscataway, 359–362.

30. Keyes, B.M., Hasoon, F., Dippo, P., Balcioglu, A., and Aboulfotuh, F. (1997), 'Influence of Na on the electro-optical properties of $Cu(In,Ga)Se_2$', Conf. Record 26th. IEEE Photovoltaic Specialists Conf., Anaheim, IEEE Press, Piscataway, 479–482.

31. Rau, U., Schmitt, M., Engelhardt, F., Seifert, O., Parisi, J., Riedl, W., Rimmasch, J., and Karg, F., 'Impact of Na and S incorporation on the electronic transport mechanisms of $Cu(In,Ga)Se_2$ solar cells', *Solid State Commun.* 107 (1998b) 59–63.

32. Schäffler, R., Hariskos, D., Kaiser, M., Ruckh, M., Rühle, U., and Schock, H.-W., 'Electrical Characterization of $Cu(In,Ga)Se_2$ Chalcopyrite Thin Films: Methods of Influencing the Effective Doping', *Crys. Res. Technol.* 31 (1996) 543.

33. Braunger, D., Hariskos, D., Bilger, G., Rau, U., and Schock, H.-W., 'Influence of Sodium on the Growth of Polycrystalline $Cu(In,Ga)Se_2$ Thin Films', *Thin Solid Films* 361–362 (2000) 161–166.

34. Braunger, D., Zweigart, S., and Schock, H.-W., 'The Influence of Na and Ga on The Incorporation of the Chalcogen in Polycrystalline $Cu(In,Ga)(S,Se)_2$ Thin-Films for Photovoltaic Applications', in Proc. 2nd World Conf. On Photovolt.

Energy Conv., edited by J. Schmidt, H. A. Ossenbrink, P. Helm, H. Ehmann, and E. D. Dunlop (E. C. Joint Res. Center, Luxembourg, 1998) 1113.

35. Wolf, D., Müller, G., 'Thin film calorimetry as a tool for in-situ investigation of reactions in the Cu-In-Se ternary system', Proceedings of the 11th International Conference on Ternary and Multinary Compounds. ICTMC-11., Institute of Physics Publishing, Bristol, UK (1998) 281.

36. Niles, D.W., Ramanathan, K., Haason, F., Noufi, R., Tielsh, B.J., and Fulghum, J.E., 'Na impurity chemistry in photovoltaic CIGS thin films: Investigation with X-ray photoelectron spectroscopy', J. Vac. Sci. Technol. A 15 (1997) 3044–3049.

37. Niles, D.W., Al-Jassim, M., and Ramanathan, K., 'Direct observation of Na and O impurities at grain surfaces of $CuInSe_2$ thin films', J. Vac. Sci. Technol. A 17 (1999) 291–296.

38. Kronik, L., Cahen, D., and Schock, H.-W., 'Effects of sodium on polycrystalline $Cu(In,Ga)Se_2$ and its solar cell performance', Adv. Mater. 10 (1998) 31–36.

39. Braunger, D., Hariskos, D. and Schock, H.W. (1998a), 'Na-related stability issues in highly efficient polycrystalline $Cu(In,Ga)Se_2$ solar cells', Proc. 2nd. World Conf. on Photovoltaic Solar Energy Conversion, Vienna, European Commission, 511–514.

40. Schroeder, D.J. and Rockett, A.A., 'Electronic effects of sodium in epitaxial $CuIn_{1-x}Ga_xSe_2$', J. Appl. Phys. 82 (1997) 5982–5985.

41. Kötschau, I.M., Turcu, M., Rau, U., Schock, H.-W., 'Structural and Electronic properties of $Cu(In,Ga)(Se,S)_2$ alloys', Mater. Res. Symp. Proc. 668 (2001) H4.5.

42. Turcu, M., Koetschau, I.M., and Rau, U., 'Composition dependence of defect energies and band alignments in the $Cu\ In_{1-x}\ Ga_xSe_{1-y}\ S_y$ alloy system', J. Apl. Phys. 91 (2002) 1391.

43. Tiwari, A.N., Blunier, S., Zogg, H., Filzmoser, M., Schmid, D., and Schock, H.-W., 'Characterization of Heteroepitaxial $CuIn_3Se_5$ and $CuInSe_2$ Layers on Si Substrates', Appl. Phys. Lett. 65/26 (1994) 3347.

44. Niki, S., Fons, P. J., Yamada, A., Suzuki, R., Ohdaira, T., Ishibashi, S., and Oyanagai, H. (1994), 'High quality $CuInSe_2$ epitaxial films-molecular beam epitaxial growth and intrinsic properties', Inst. Phys Conf. Ser. 152E (1994) 21–227.

45. Klenk, R., Walter, T., Schock, H.-W., and Cahen, D., 'A model for the successful growth of polycrystalline films of $CuInSe_2$ by multisource physical vacuum evaporation', Adv. Mater. 5 (1993) 114–119.

46. Probst, V., Karg, F., Rimmasch, J., Riedl, W., Stetter, W., Harms, H., and Eibl, O., 'Advanced stacked elemental layer progress for $Cu(InGa)Se_2$ thin film photovoltaic devices', Mater. Res. Soc. Symp. Proc. 426 (1996) 165–176.

47. Kazmerski, L.L., Ayyagari, M.S., White, F.R., Sanborn, G.A., 'Growth and Properties of vacuum deposited $CuInSe_2$ thin films', J. Vac. Sci. Technol. 13 (1976) 139.

48. Zweigart, S., Schock, H.-W., and Powalla, M., 'A New Method for the Analysis of Film Formation Kinetics and a Simple Process for the Growth of $Cu(In,Ga)Se_2$', in Proc. 14th Europ. Photovolt. Solar Energy Conf., edited by H.A. Ossenbrink, P. Helm, and H. Ehmann (H.S. Stephens & Ass., Bedford, UK, 1997) 1254.

49. Wada, T., Nishikawa, S., Hashimoto, Y., and Negami, T., 'Physical vapour deposition of $Cu(In,GaI)Se_2$ Films for Industrial Applications', Mater. Res. Symp. Soc. Proc. 668 (2001) H.2.1.1.

50. Mickelsen, R.A., and Chen, W.S., 'High photocurrent polycrystalline thin-film CdS/CuInSe$_2$ solar cell', *Appl. Phys. Lett.* 36 (1980) 371–373.
51. Mickelsen, R.A., and Chen, W.S. (1982), 'Polycrystalline thin-film CuInSe$_2$ solar cells', Conf. Record 16th. IEEE Photovoltaic Specialists Conf., San Diego, IEEE Press, Piscataway, 781–785.
52. Kessler, J., Schmid, D., Dittrich, H., and Schock, H.-W., 'CuInSe$_2$ Film Formation from Sequential Deposition of In(Se):Cu:Se', in Proc. 12th European Photovoltaic Solar Energy Conference (Amsterdam, 1994) 648.
53. Gabor, A.M., Tuttle, J.R., Albin, D.S., Contreras, M.A., Noufi, R., and Hermann, A.M., 'High-efficiency CuInxGa$_{1-x}$Se$_2$ solar cells from (InxGa$_{1-x}$)$_2$Se$_3$ precursors', *Appl. Phys. Lett.* 65 (1994) 198–200.
54. Gabor, A.M., Tuttle, J.R., Bode, M.H., Franz, A., Tennant, A.L., Contreras, M.A., Noufi, R., Jensen, D.G., and Hermann, A.M, 'Band-gap engineering in Cu(In,Ga)Se$_2$ thin films grown from (In,Ga)$_2$Se$_3$ precursors', *Solar Energy Mater. Solar Cells* 41 (1996) 247–260.
55. Dullweber, T., Hanna, T., Rau, U., and Schock, H.-W., 'A New Approach to High Efficiency Solar Cells by Band Gap Grading in Cu(In,Ga)Se$_2$ Chalcopyrite Semiconductors', *Solar Energy Mater. Solar Cells* 67 (2001) 145–150.
56. Dimmler, B., and Schock, H.-W., 'Scaling-up of CIS Technology for Thin-film Solar Modules', *Progr. Photovolt. Res. Applic.* 4 (1996) 425.
57. Binsma, J.J.M., and Van der Linden, H.A., 'Preparation of thin CuInS$_2$ films via a two-stage process', *Thin Solid Films* 97 (1982) 237–243.
58. Chu, T.L., Chu, S.C., Lin, S.C., and Yue, J., 'Large grain copper indium diselenide films', *J. Electrochem. Soc.* 131 (1984) 2182–2184.
59. Kapur, V.K., Basol, B.M., and Tseng, E.S., 'Low-cost methods for the production of semiconductor films for CuInSe$_2$/CdS solar cells', *Solar Cells* 21 (1987) 65–72.
60. Mitchell, K.C.E., Ermer, J., and Pier, D. (1988), 'Single and tandem junction CuInSe$_2$ cell and module technology', Conf. Record 20th. IEEE Photovoltaic Specialists Conf., Las Vegas, IEEE Press, Piscataway, 1384–1389.
61. Gallon, P.N.R., Orsal, G., Artaud, M.C., and Duchemin, S. (1998), 'Studies of CuInSe$_2$ and CuGaSe$_2$ thin films grown by MOCVD from three organometallic sources', Proc. 2nd. World Conf. on Photovoltaic Solar Energy Conversion, Vienna, European Commission, 515–518.
62. Fischer, D., Dylla, T., Meyer, N., Jäger-Waldau, A., Lux-Steiner, M-Ch., 'CVD for thin film solar cells employing two binary sources', *Thin Solid Films* 387 (2001) 63–66.
63. Thouin, L., and Vedel, J. *J. Electrochemical Soc.* 142 (1995) 2995.
64. Abken, A., Heinemann, F., Kampmann, A., Leinkühler, G., Rechid, J., Sittinger, V., Wietler, T. and Reineke-Koch, R. (1998), 'Large area electrodeposition of Cu(In,Ga)Se$_2$ precursors for the fabrication of thin film solar cells', Proc. 2nd. World Conf. on Photovoltaic Solar Energy Conversion, Vienna, European Commission, 1133–1136.
65. Guillien, C., and Herrero, J., *Thin Solid Films*, 387 (2001) 57–59.
66. Eberspacher, C., Pauls, K.L., and Fredric, C.V. (1998), 'Improved processes for forming CuInSe$_2$ films', Proc. 2nd. World Conf. on Photovoltaic Solar Energy Conversion, Vienna, European Commission, 303–306.
67. Kapur, V.K., Fisher, M., and Roe, R., *Mater. Res. Symp. Proc.* 668 (2001) H2.6.

68. Ramanathan, K., Bhattacharya, R.N., Granata, J., Webb, J., Niles, D., Contreras, M.A., Wiesner, H., Haason, F.S., and Noufi, R. (1998a), 'Advances in CIGS solar cell research at NREL', Conf. Record 26th. IEEE Photovoltaic Specialists Conf., Anaheim, IEEE Press, Piscataway, 319–325.
69. Battacharya, R.N., and Fernandez, A.M., *Mater. Res. Symp. Proc.* 668 (2001) H8.10.

Development of Cu(InGa)Se$_2$ Thin-Film Solar Cells

Makoto Konagai and Katsumi Kushiya

11.1 High-Efficiency Techniques

11.1.1 Efficiencies of Small-Area CIGS Thin-Film Solar Cells

Cu(InGa)Se$_2$(CIGS)-based-material systems are attractive for photovoltaic device applications because they have shown very good stability in outdoor tests, which is an important criterion for commercialization. The performance of CIGS thin-film solar cells has improved markedly since the Boeing group reported the achievement of over 10% by using evaporation techniques [1].

At present, CIGS absorber layers are deposited by a three-stage evaporation or a selenization method. Up to now, several groups have reported high efficiencies of over 18% [2] and more than 12% [3] in the case of modules. These device structures essentially consisted of ZnO/CdS/CIGS/Mo/glass as shown in Fig. 11.1. The CIGS thin-film solar cells consist of several thin layers and interfaces. In addition to improvements in the quality of the CIGS

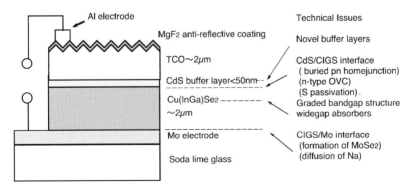

Fig. 11.1. Structure of a CIGS thin-film solar cell and issues to improve the efficiency

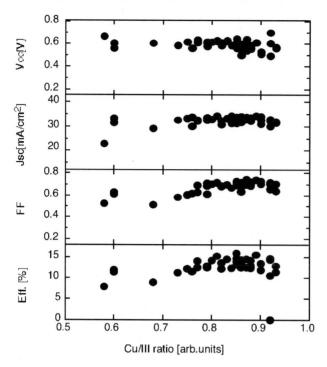

Fig. 11.2. Photovoltaic performances of CIGS thin-film solar cells as a function of Cu/III ratio

absorber layer, the properties at the CIGS/Mo and CdS/CIGS interfaces affect the device performance. Technical issues that have a major effect on the photovoltaic performances are also shown in Fig. 11.1.

In CIGS thin-film solar cells, soda lime glass is widely used as a substrate. It is reported that the Na diffusion from the soda lime substrate into the CIGS film enhances the grain growth of the absorber layer and is also effective to increase the hole concentration. Na also can be intentionally doped by using Na_2O_2, Na_2S, Na_2Se, and NaF etc.

One of the important parameters to obtain high efficiencies in CIGS-based solar cells is the Cu/III(In+Ga) ratio in the film. As shown in Fig. 11.2, high efficiencies of over 10% are obtained in a relatively wide range of Cu/III ratios from 0.75 to 0.95. Most devices made of CIGS absorbers with compositions in the range of higher Cu/III ratio(> 0.95) do not show any photovoltaic effect. The solar cells are shorted due to undesirable Cu-related secondary phases with significantly lower resistivities.

The Ga composition ratio is typically $0.2 \sim 0.3$. Further increase of the Ga composition ratio usually degrades the cell efficiency. In three-stage evaporation, the graded-bandgap structure as shown in Fig. 11.3 is unintentionally prepared during the deposition. This band structure improves the effective

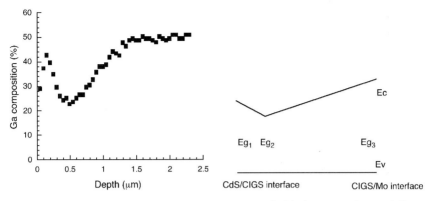

Fig. 11.3. Bandgap profiling of CIGS solar cell: (left) Auger analysis of Ga composition, (right) graded-bandgap structure

hole diffusion length in the absorber layer, thus the internal recombination loss in the visible region becomes negligible. The open-circuit voltage of CIGS thin-film solar cells is roughly expressed by the following equation: $V_{oc} = E_g/e - 0.5$ (V) for a Ga composition ratio smaller than 0.3, where E_g is the bandgap of CIGS. However, the open-circuit voltage for Ga compositions above 0.4 becomes lower than expected. The reason for this is partly the degradation of film quality and type conversion of the surface layer from n-type to p-type. In the conventional CIGS cells, the surface layer is n-type because of the diffusion of Cd during the CBD process of CdS deposition and also the formation of n-type Cu(InGa)₃Se₅ (OVC, ordered vacancy phases). It is reported that the surface layer of CIGS with compositions above 0.3 become highly resistive or p-type due to the type conversion of Cu(InGa)₃Se₅ layer. A sulfurization process is also applied to the surface of CIGS to convert the surface Cu(InGa)Se₂ to Cu(In,Ga)(Se,S)₂. The addition of S to the surface layer improves the photovoltaic performance of the cell. The role of S is passivation of Se vacancies and dangling bonds.

A MoSe₂ layer is formed at the CIGS and Mo interface [4, 5]. It is reported that a columnar structure MoSe₂ layer exists at the CIGS/Mo interface in a high-efficiency CIGS solar cell and that the c-axis of MoSe₂ is oriented parallel to the surface of Mo layer. Due to the layer structure of the MoSe₂ compound, if the MoSe₂ grains are oriented perpendicular to the surface of the Mo layer, the CIGS film would tend to delaminate from the Mo layer due to ease of interlayer cleaving. Thus, crystal orientation of the MoSe₂ grains is important for the fabrication of solar cells.

Figures 11.4 and 11.5 show typical I–V characteristics and spectral response of CdS/CIGS thin-film solar cells, respectively. The high-efficiency CIGS solar cells show an open-circuit voltage of 0.670–0.680 V, a short-circuit current of 35.0–36.0 mA cm⁻² and fill factor of 0.76–0.79, respectively. As shown in Fig. 11.5, the internal recombination loss is very small for these devices.

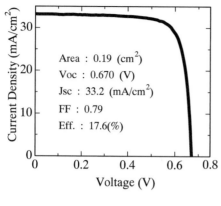

Fig. 11.4. Illuminated I–V characteristics of a high-efficiency CdS/CIGS thin-film solar cell

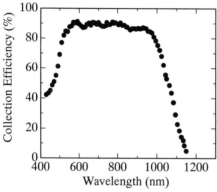

Fig. 11.5. Typical collection efficiency spectra of CdS/CIGS thin-film solar cell

11.1.2 In-situ Monitoring of Composition Ratio

The composition ratio of the CIGS films has a great influence on the performance of the cell. Thus the development of in-situ monitoring techniques of the composition ratio of the growing surface is quite important for improving reproducibility. A major breakthrough has been achieved by in-situ monitoring of the growing surface temperature. Kohara et al. [6, 7] found that the temperature of growing CIGS surface was dependent on the composition ratio of Cu/(In+Ga) when the film was deposited under constant heating power. The composition-monitoring system was applied to a three-stage deposition process of the CIGS films. Figure 11.6 shows a typical deposition sequence of the three-stage process and the in-situ substrate temperature variation. In the first stage, In-Ga-Se precursors were deposited on a Mo-coated soda lime glass. Then, Cu and Se fluxes were exposed to the In-Ga-Se precursors to form Cu-rich CIGS films in the second stage. Finally, small amounts of In, Ga, and Se fluxes are added to the Cu-rich CIGS films in order to obtain slightly (In,Ga)-rich CIGS films. Figure 11.6 (middle) is a typical variation of the substrate temperature as a function of the deposition time during a part of the second and the third stages under constant heating power. The chem-

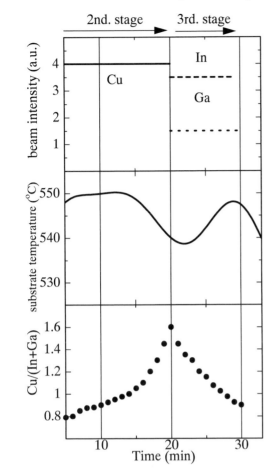

Fig. 11.6. Typical variation of (top) beam fluxes of Cu, In, and Ga, (middle) temperature of the growing film, and (bottom) Cu/(In+Ga) ratios of the films removed at each point [6]

ical composition of the films determined by EDX (energy dispersive X-ray analysis) is also shown in Fig. 11.6 (bottom).

These results suggest that when the Cu/(In+Ga) ratio of the growing film is less than one, the film temperature is constant at around 550°C, and when the Cu/(In+Ga) ratio is above one, the film temperature becomes lower than 550°C, even under constant heating power. For the Cu-rich films, the temperature of the growing film decreases with the Cu/(In+Ga) ratio. This effect strongly indicates that the Cu/(In+Ga) ratio of the growing CIGS film can be detected by measuring the film-temperature variation. The temperature variation of the growing film can be explained by the dependence of emissivity on the Cu/(In+Ga) ratio. The In(Ga)-rich CIGS phases have similar emissivity

Table 11.1. Solar-cell parameters for the CIGS cells with efficiencies of over and below 18%

Cell number	CIS520	CIS1034
Efficiency (%)	17.6	18.5
Open-circuit voltage V_{oc} (V)	0.649	0.674
Short-circuit current I_{sc} (mA cm^{-2})	36.1	35.4
Fill factor	0.751	0.774
Series resistance (Ω cm^{-2})	0.5	0.5
Shunt resistance (Ω cm^{-2})	7400	4800
Diode quality factor n	1.5	1.4
Saturation current I_0 (A cm^{-2})	7×10^{-9}	6×10^{-10}

to the stoichiometric CIGS and the Cu-rich CIGS in which the Cu$_x$Se phase exists at the surface has a higher emissivity than the stoichiometric CIGS. A cell with an active area of 0.96 cm^2 showed an efficiency of 17.6%, as shown in Table 11.1 by using this technique.

In addition to the precise control of the composition of the absorber layer, the technique to deposit CdS layer very uniformly by the chemical-bath deposition technique has been developed. As the process is improved in order to prepare CdS film with excellent uniformity, damage and/or defects at the surface of the CIGS films, possibly formed by sputtering of ZnO films, may decrease due to sufficient coverage of CdS films over CIGS surface.

A schematic explanation is shown in Fig. 11.7 [8]. In this structure, Cd could uniformly diffuse into the CIGS films and a uniform buried junction may decrease the recombination at the interface of the pn junction, apart from the CIGS surface. Up to now, an efficiency of 18.5% has been obtained, as shown in Table 11.1 under AM1.5 illumination. The saturation current density I_0 of the cell of 18.5% efficiency is one-tenth that of the cell with 17.6% efficiency.

11.1.3 Buffer Layers

11.1.3.1 CdS

The CdS films deposited by CBD process are widely used as an interfacial layer of the CIGS thin-film solar cells in order to improve the efficiency. The role of the CBD-CdS buffer layer has been thought of as a prevention of undesirable shunt paths through the portion of the very thin CdS buffer layer and the protection of the junction region from sputtering damage during subsequent transparent conducting oxide (TCO) deposition.

Damage and/or defect

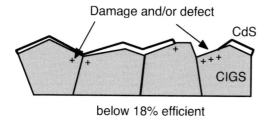

below 18% efficient

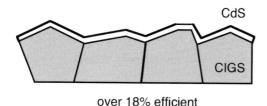

over 18% efficient

Fig. 11.7. Schematic drawing of the sufficient coverage of CdS films over CIGS surface [8]

Recently, several research groups reported that the role of CBD-CdS is the diffusion of Cd into the CIGS absorber and the formation of a buried pn junction inside the absorber [9]. Wada et al. [10] reported that when CIGS films were soaked in the Cd^{2+} aqueous solution, Cd atoms were doped in the CIGS surface by the substitution of Cu atoms in the CIGS film by Cd ions in the solution. SIMS depth profiles clearly demonstrated that the Cd concentration is the highest at the CIGS surface and it decreases with depth into the CIGS films. The Cd concentration reached the background level at a depth of about 20 nm. They also observed a decrease of Cu concentration near the film surface. The decrease of Cu concentration starts at depth of 20 nm, which is in good agreement with the limit depth of Cd doping. Cd atoms are doped in the CIGS surface by substitution reaction such as,

$$x\mathrm{Cd}^{2+} + \mathrm{Cu(InGa)Se}_2 \longrightarrow (\mathrm{Cd}_x\mathrm{Cu}_{1-2x})(\mathrm{InGa})\mathrm{Se}_2 + 2x\mathrm{Cu}^+ \ .$$

They also demonstrated that Cd is more easily doped in the CuIn$_3$Se$_5$ than in the CuInSe$_2$. This indicates that CuIn$_3$Se$_5$ phase on the physical-vapor-deposited CIGS is readily modified during the CBD process of the Cd layer.

11.1.3.2 Zn Compound Buffer Layer

In order to improve the feasibility of manufacturing and the environmental safety, it is desirable to replace CdS with an alternative buffer material. Kushiya et al. [3] proposed a Zn compound that contains sulfur as one alternative for Cd-free buffers. A Zn-compound buffer layer was deposited on the CIGS thin-film absorber by the CBD method using Zn salts, S compounds, and ammonia solution. A submodule efficiency of 12% was reported for a device structure consisting of ZnO/Zn(O,S,OH)$_x$/CIGS/Mo on a soda-lime glass substrate [3]. More details are described in Sect. 11.2.

11.1.3.3 ZnS

In general, the diffusion of zinc into CIGS thin-film is more difficult than the case of Cd. Some external driving force is required to enhance the diffusion of Zn into CIGS thin-films to form a buried pn homojunction. Nakada and Mizutuni [11] demonstrated a role of the annealing process in CBD-ZnS/CIGS thin-film solar cells for preparing the buried homojunction. CBD-ZnS layers were chemically grown on CIGS thin films using a $ZnSO_4$-ammonia-thiourea aqueous solution at 80°C. CBD-ZnS thin films of 20–40 nm thick were grown after dipping CIGS thin films for 15 min. The samples were treated at elevated temperatures in air. Following this, ZnS layers were removed by hydrochloric acid. The diffusion behavior at the CBD-ZnS/CIGS interface was investigated by means of SIMS and EDX. It was found that Zn diffused into CIGS thin films after heat treatment even at relatively low temperature of 200°C for 10 min. The heat treatment promotes the Zn diffusion into the CIGS thin film. EDX analysis was carried out for ZnO/ZnS/CIGS solar-cell structures to confirm the Zn diffusion by air annealing. Zn was present inside the CIGS layer at a position of 40–50 nm from the ZnS/CIGS interface boundary. Thus it is evident that the buried homojunction is formed at the surface of the CIGS layer by using Zn.

Figure 11.8 shows a cell structure of CIGS thin-film solar cells with the MgF_2/ZnO:Al/CIGS/Mo/soda lime glass structure along with a cross-sectional transmission electron micrograph (TEM). The absorber layer is deposited by the the three-stage process. A 130–150 nm-thick CBD-ZnS buffer layer was deposited on the CIGS absorber layer. Cell performances improved markedly as the thickness of the CBD-ZnS buffer layer was increased to 130 nm. Up to now, the best efficiency of 17.7% was obtained with a CBD-ZnS buffer layer (Table 11.2).

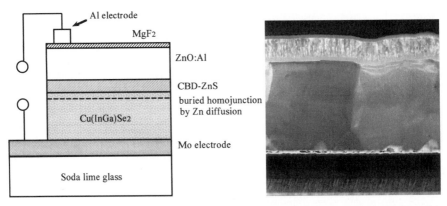

Fig. 11.8. (left) Cell structure of a CIGS thin-film solar cell with CBD-ZnS buffer layer and (right) a cross sectional TEM image [11]

Table 11.2. Photovoltaic performances of small-area Cu(InGa)Se$_2$ and CuInS$_2$ thin-film solar cells with a CdS buffer layer and a novel buffer layer (η^*: Efficiency, **: Tokyo Institute of Technology, CIGS: Cu(InGa)Se$_2$)

Structure	$V(V)$	I_{sc} (mA/cm^2)	FF	$\eta*$ (%)	Area (cm^2)	Reasearch Institute
ZnO/CdS/Cu(InGa)Se$_2$	0.678	35.22	0.7865	18.8	0.449	NREL
ITO/ZnO/CdS/CIGS	0.674	35.4	0.774	18.5	0.96	Matsushita
ZnO:B/ZnO/CdS/CIGS	0.645	36.8	0.760	18.0	0.20	Aoyama Univ.
ZnO/CdS/CIGS	0.671	33.2	0.790	17.6	0.19	TIT**
ZnO/CdS/CIGS	0.647	35.8	0.760	17.6	0.38	EC group
ZnO/ZnS/CIGS	0.671	34.0	0.776	17.7	0.15	Aoyama Univ.
ZnO/Zn(SeOH) /Cu(InGa)(SeS)$_2$	0.566	33.0	0.701	15.1	1.08	HMI/Siemens
ZnO/In(OH,S)CIGS	0.594	35.5	0.746	15.7	0.38	Stuttgart Univ.
ZnO/ZnInSe/CIGS	0.652	30.4	0.763	15.1	0.19	TIT
ZnO:B/ALD-ZnO/CIGS	0.578	35.3	0.713	14.6	0.19	TIT
Zn$_{1-x}$Mg$_x$O/CIGS	0.572	33.8	0.682	13.2	0.96	Matsushita
ZnO/CdS/CuInS$_2$	0.715	23.7	0.71	12.0	-	Stuttgart Univ.
ZnO/CdS/CuInS$_2$	0.728	21.42	0.71	11.1	0.48	HMI

11.1.3.4 ZnIn$_x$Se$_y$

In$_x$Se$_y$ (IS) and ZnIn$_x$Se$_y$ (ZIS) buffer layers have been employed as attractive alternatives to the CdS buffer layer [12]. These buffer layers were continuously coevaporated on a CIGS absorber layer in the same fabrication apparatus. This continuous fabrication process makes it possible to fabricate the CIGS thin-film solar cells by all-dry processes. These buffer layers have the following advantages compared to the CBD-CdS buffer layer:

- The same elemental effusion cells for the CIGS absorber layers can be used for the IS, resulting in a reduction of the cost for the CIGS solar cells.
- ZIS has a defect chalcopyrite structure in the form of ZnIn$_2$Se$_4$, similar to the chalcopyrite structure of CIGS. Therefore, this buffer layer is suitable for the fabrication of a good heterointerface with the CIGS absorber.

ZIS/CIGS thin-film solar cells were fabricated by three-stage evaporation. After the deposition of the absorber layer, the substrate was cooled to the growth temperature of the buffer layer. The IS or the ZIS buffer layer was deposited continuously on the CIGS absorber layer by a coevaporation method at a substrate temperature of about 300°C. The Se/Zn ratio and the Se/In ratio were adjusted to 10 and 15, respectively. The structural and electrical properties of ZIS thin films were characterized. It was demonstrated by the X-ray diffraction patterns and the Raman spectroscopy measurements that polycrystalline ZIS thin films were grown by the coevaporation method. The bandgap of ZIS is estimated at around 2.0 eV. In addition, the ZIS films on glass substrates showed low dark conductivity of 10^{-9}–10^{-8} S cm^{-1} and high photosensitivity of 10^3–10^4. The preferential growth of ZnIn$_2$Se$_4$ with (112) lattice plane was observed at a higher temperature range of 400–500°C.

Based upon the above fundamental results, CIGS thin-film solar cells with a ZnO/ZIS/CIGS structure were fabricated. Table 11.2 summarizes the efficiencies of these solar cells. The best efficiency is 15.1% with a remarkable improvement in fill factor (0.763). So far, a little lower efficiencies of CIGS solar cells with ZIS buffer layers were mainly attributed to lower short-circuit currents. A reduction of the short-wavelength response due to the absorption of the ZIS buffer layer is observed.

11.1.3.5 ZnO

A high-resistivity ZnO layer grown by the atomic layer deposition (ALD) method has been studied, especially for application as a buffer layer in CIGS-based solar cells [13]. ALD is performed under vacuum that can be incorporated into inline manufacturing. Moreover, due to its surface-controlled nature, the ALD technique offers a possibility to deposit reproducible and uniform film over large areas along with better control of thickness. To investigate the optimum growth parameters for a high-resistivity ZnO layer, undoped ZnO films were deposited by ALD on glass substrate at temperature of 165°C. Diethylzinc (DEZn) and H$_2$O were used as the zinc and oxygen source, respectively. These source gases were alternately fed into the chamber with argon (Ar) carrier gas. The periodic switching of the air valves was automatically controlled by a personal computer. The typical pulse lengths used were 2 s for the reactants and 8 s for the evacuation between the reactants.

Figure 11.9 shows the dependence of resistivity on the DEZn flow rate. The deposition rate varied from 0.3–0.8 nm/cycle, depending on the DEZn flow rate. It can be seen from Fig. 11.9 that the resistivity of the films is strongly dependent on the DEZn flow rate. Although, in the DEZn flow rate range

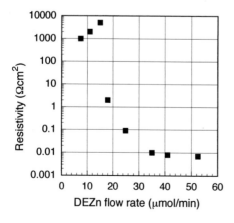

Fig. 11.9. Resistivity of undoped ALD-ZnO films as a function of the DEZn flow rate

of 40–53 μmol min^{-1}, the average growth rate led to almost a monolayer growth, the resistivity of the films was lower than 10^{-2} Ω cm. These low-resistive films could not be applied as buffer layers in CIGS solar cells. The resistivity increased rapidly with decreasing DEZn flow rate, but subsequently decreased slightly with further decrease in the DEZn flow rate. The increase in resistivity with decreasing DEZn flow rate can be attributed to the decrease in defects such as oxygen vacancies and/or interstitial zinc acting as donors in the film. A high-resistivity ZnO ($\sim$ 1 kΩ cm) was obtained for the films prepared between the flow rates of 8 and 14.5 μmol min^{-1}. These results indicate that a high-resistivity ZnO film prepared by ALD can be controlled by varying the DEZn flow rate at a constant H$_2$O flow rate.

In order to study ALD-ZnO buffers, a number of devices with the ZnO:B/i-ZnO/CIGS/Mo/glass structure were fabricated. The Cu(InGa)Se$_2$ absorbers with thin Cu(InGa)(SeS)$_2$ surface layers were fabricated by selenization and sulfurization, yielding a homogeneous composition over a 10×10 cm^2 area. Details on the fabrication of CIGS absorber layers can be found in Sect. 11.2.

Figure 11.10 shows the solar-cell efficiencies as a function of resistivity of undoped ZnO buffer layer. The thickness of the buffer layer was set at 75 nm. By using ZnO buffer layer, a relatively high short-current density of around 35.5 mA cm^{-2} was achieved because of the elimination of CdS-related absorption losses. The short-circuit current density does not exactly depend on the buffer-layer resistivity, whereas open-circuit voltage and fill factor are significantly dependent on the buffer layer resistivity. The high-efficiency solar cells were fabricated by using ZnO with resistivities higher than 10^3 Ω cm. Up to now, an efficiency of 14.6% has been obtained by using ALD-ZnO as shown in Table 11.2.

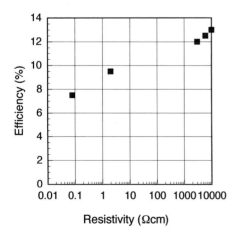

Fig. 11.10. Photovoltaic performances of ZnO/CIGS solar cells as a function of resistivity of undoped ZnO buffer layer

11.1.4 Conduction Band Offset

Various groups have proposed Cd-free window layers, but none of the CIGS solar cells developed with Cd-free window layers have achieved a higher efficiency than those with CdS. One of the important factors in improving the efficiency of solar cells is the formation of an appropriate conduction-band offset between the CdS and CIGS layers. Negami et al. [14] reported the effects of conduction-band offset of window/CIGS layers on solar-cell performance by using $Zn_{1-x}Mg_xO$. $Zn_{1-x}Mg_xO$ thin films were prepared using a cosputtering method from ZnO and MgO targets at room temperature. The $Zn_{1-x}Mg_xO$ films were deposited on glass substrates and directly on a CIGS film to characterize bandgap and valence-band offsets (VBO), respectively. The bandgaps of the $Zn_{1-x}Mg_xO$ films were derived from optical absorption data, and the VBO between the $Zn_{1-x}Mg_xO$ and CIGS films was characterized by X-ray photoelectron spectroscopy (XPS). Figure 11.11 shows the band alignment of the CIGS and $Zn_{1-x}Mg_xO$ films with the different Mg content. The valence band of $Zn_{1-x}Mg_xO$ was constant except for the $Zn_{0.83}Mg_{0.17}O$ film. The conduction band of $Zn_{1-x}Mg_xO$ shifted to the vacuum level mainly due to an increase of the bandgap. These results indicate that the conduction-band offset of $Zn_{1-x}Mg_xO$/CIGS films can be controlled by the Mg content of the $Zn_{1-x}Mg_xO$ film.

CIGS solar cells with a structure of ITO/$Zn_{1-x}Mg_xO$/CIGS/Mo/glass were fabricated. CIGS thin films were deposited by using the three-stage evaporation. Then, the CIGS films were dipped in the solution with Cd^{2+} ion to prepare Cd-doped CIGS layers and to form a buried junction. The thickness of the Cd-doped layer was estimated to be less than 0.02 μm by a SIMS profile. The CIGS solar cells with the different conduction-band offsets of the $Zn_{1-x}Mg_xO$/CIGS layers were fabricated by changing the Mg content of the $Zn_{1-x}Mg_xO$ films. The bandgap of the CIGS films was fixed. Figure 11.12 shows the efficiency as a function of conduction-band offset of $Zn_{1-x}Mg_xO$/CIGS films. The positive band offset of conduction bands indi-

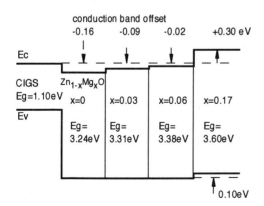

Fig. 11.11. Band alignment of CIGS and $Zn_{1-x}Mg_xO$ with different Mg contents [14]

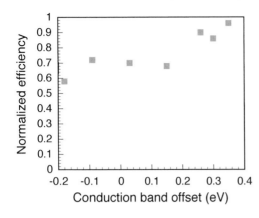

Fig. 11.12. Normalized efficiency as a function of conduction-band offset [14]

cates that the conduction-band edge of the $Zn_{1-x}Mg_xO$ film is above that of the CIGS film. The efficiency in Fig. 11.12 is normalized to that of the typical solar cell that consists of the CdS buffer layer on the same CIGS film. An efficiency of the cell with the conduction-band offset above 0.25 eV increased abruptly up to almost the same level of the base-line solar cell. According to the theoretical analysis, the degradation of efficiencies for the band offset lower than 0.2 eV is due to the recombination between the majority carriers via defects at the window/CIGS interface. The highest efficiency of the $Zn_{1-x}Mg_xO$/CIGS cells without an antireflective coating was 13.2% ($V_{oc} = 0.572$ V, $I_{sc} = 33.8$ mA cm^{-2}, FF $= 0.682$, 0.96 cm^2).

11.1.5 Flexible Substrates

There is interest in the development of CIGS solar cells on flexible substrates for novel applications to space, buildings, etc. Several groups have reported on the flexible CIGS solar cells. Stainless steel can be essentially applicable to the high-temperature process. Stainless steel has attracted attention as a substrate for the high-efficiency flexible CIGS solar cells. Up to now, an efficiency exceeding 17% has been obtained using a Mo-coated stainless steel substrate [2]. However, in order to develop the large-area modules, an insulating layers between Mo back electrodes and stainless steel substrates is needed for interconnection. Figure 11.13 show a CIGS solar-cell structure on a stainless steel substrate with an insulating layer [15]. Stainless steel with a thickness less than 0.1 mm was used as a substrate. The insulating layer of SiO$_2$ or Al$_2$O$_3$ with a thickness of 300 nm and 150 nm, respectively, was deposited by RF magnetron sputtering directly on the stainless steel substrate. The thickness of Mo back electrodes is less than 1 μm. The CIGS absorber layer was prepared by the three-stage process at the highest temperature of about 500°C. Table 11.3 summarizes the cell performances for samples with and without the insulating layer. For CIGS solar cells with the insulating layer, an efficiency of 11.0% was obtained without an antireflective coating. The efficiency scarcely

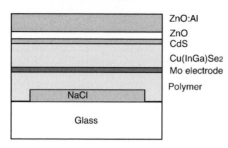

Fig. 11.13. Schematic diagram of the CdS/CIGS solar cell on a polymer-coated glass substrate with a NaCl buffer layer. The glass substrate is removed by dissolving the NaCl buffer layer and the cells are obtained on 20–25 μm-thick polymer sheets [17]

decreased by the insertion of SiO_2 layer between the Mo film and stainless steel substrate, however the V_{oc} of cells deposited on stainless steel is lower than on soda lime glass. With an Al_2O_3 insulating layer, instead of SiO_2, the same efficiency was obtained. The CIGS films prepared on the stainless steel do not contain Na. Further improvement of the cell performance can be obtained by Na-doping.

The effect of Na addition on the cell performance of flexible solar cells was also investigated. Na for the absorber was provided by NaF precursor films deposited prior to CIGS coevaporation [16]. It was demonstrated that adding a 3–10 nm NaF precursor layer improves the efficiencies on polymer substrate and Ti foils. The optimum precursor layer thickness was about 10–20 nm for CIGS on a metal substrate. Up to now, an efficiency of 10.6% has been reported for a 0.5-cm^2 solar cell on a Ti-foil substrate.

Although a polyimide sheet is a strong candidate for the flexible-substrate, the polyimide sheet is difficult to heat above 400°C, whereas high-quality CIGS films are grown at substrate temperatures above 500°C. Thermal expansion mismatch is another problem that may cause cracks in the Mo or CIGS layers. Tiwari et al. [17] developed a novel process to obtain CIGS solar cells on polymer sheets. Figure 11.13 shows the schematic diagram of the

Table 11.3. CIGS solar-cell performances prepared on stainless steel substrate without Na doping [15]

Substrate	Stainless steel with SiO_2 insulating layer	Stainless steel without insulating layer	Soda lime glass
Area (cm^2)	0.24	0.24	0.96
V_{oc} (V)	.0527	0.516	0.590
I_{sc} (mA cm^{-2})	31.4	30.6	31.5
FF	0.664	0.685	0.732
Efficiency (%)	11.0	10.8	13.2

Table 11.4. Efficiencies of small-area CIGS solar cells prepared on flexible substrates

Structure	V_{oc} (V)	I_{sc} (mA cm^{-2})	FF	Efficiency (%)	Reference
Mo/stainless steel	0.646	36.38	0.7419	17.4	2
SiO$_2$Mo/stainless steel	0.527	31.4	0.664	11.0	15
polyimide	0.6127	30.62	0.682	12.8	17
Flexible metal	0.6668	40.83	0.763	15.2(AM0)	17

layers and substrate. First, a thin buffer layer of NaCl is evaporated on the glass substrate. Then, a polyimide is spin coated and cured at a temperature of about 400°C in air. The typical thickness of the polyimide is about 20 μm. The NaCl layer is used as a buffer because it can be dissolved easily in water and the glass substrate can be separated from the polymer film. This buffer layer can provide Na to the CIGS layer during the deposition. The CIGS absorber layer of about 2 μm was grown by coevaporation at a substrate temperature of 450°C. An efficiency of 12.8% (total area, without AR coating) has been obtained, as shown in Table 11.4.

CIGS solar cells deposited on light-weight and flexible substrates are very promising for space applications. A benefit of the use of thin-film CIGS is its radiation tolerance. CIGS has been shown to retain much more of its initial power than a comparable III–V cell. In fact, little or no degradation has been measured for CIGS thin-film solar cells due to 1 MeV electron irradiation at a fluence of 10^{16} [18]. Up to now, a 15.2% AM0 cell performance has been achieved for a cell prepared on a flexible metal substrate as shown in Table 11.4, resulting in a specific power of 1433 W kg^{-1} and 1235 W kg^{-1} for bare and SiO$_x$ coverglass cells, respectively.

11.2 Fabrication Technologies of Large-area CIGS-based Modules

11.2.1 Introduction

For all of the thin-film photovoltaic (PV) technologies, the most important issue is to improve the cost competitiveness against the crystalline Si solar cells and modules, which are currently entering an over 100 MW/y production stage in the leading companies, for instance, Kyocera and Sharp in Japan. It is acknowledged that the simplest way to reduce the cost is to increase the efficiency and among the thin-film PV technologies, Cu(InGa)Se$_2$(CIGS)-based solar cells and modules have the highest potential to achieve the competitive efficiency with crystalline Si solar cells and modules as described in Sect. 11.1.

198   Makoto Konagai and Katsumi Kushiya

Table 11.5. Groups possible to make a CIGS-based module with the efficiency of over 12% on a substrate size of 30 cm × 30 cm or larger

Organization	Device Structure /Substrate=Soda-lime glass	Efficiency (Output)	Area [cm^2]
Showa Shell Sekiyu K.K.	ZnO/ Zn(O,S,OH)$_x$/ Cu(InGa)(SeS)$_2$(CIGSS)/ CIGS/ Mo	13.4 % * (46.4 W$_p$)	3459 (Mosaic Module)
Shell Solar GmbH	ZnO/ CdS/ CIGSS/ Mo	13.1% (64.8 W$_p$)	5400 (Module)
Shell Solar Industries	ZnO/ CdS/ CIGSS/ Mo	12.8% * (46.5 W$_p$)	3626 (Module)
Würth Solar	ZnO/ CdS/ CIGS/ Mo	12.5% (74 W$_p$)	5932 (Circuit)
Showa Shell Sekiyu K.K..	ZnO/ Zn(O,S,OH)$_x$/ CIGSS/ CIGS/ Mo	14.2 % (12.3 W$_p$)	864 (Circuit)
ZSW GmbH	ZnO/ CdS/ CIGS/ Mo	12.7 % # (9.3 W$_p$)	736 (Circuit)

Measurements: *   National   Renewable   Energy   Laboratory   (NREL), # Fraunhofer-Institut für Solare Energiesysteme (FhG/ISE)

Up to now, three groups (Siemens Solar Industries (currently Shell Solar Industries), ZSW (Zentrum für Sonnenenergie- und Wasserstoff-Forschung, or Center for Solar Energy and Hydrogen Research) GmbH, and Showa Shell

Fig. 11.14. ST-series CIGS-based modules in Shell Solar Industries

Crystalline Si solar modules   **CIGS-based modules**

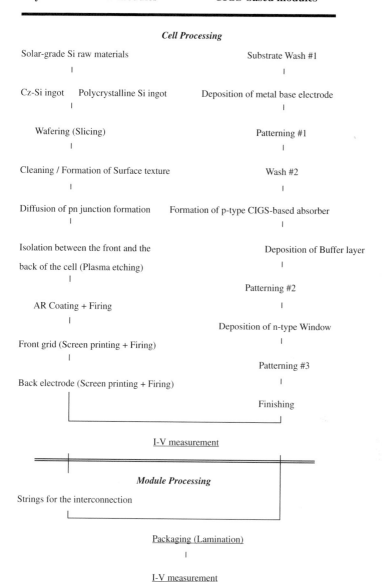

Cell Processing

Crystalline Si	CIGS-based
Solar-grade Si raw materials	Substrate Wash #1
Cz-Si ingot Polycrystalline Si ingot	Deposition of metal base electrode
Wafering (Slicing)	Patterning #1
Cleaning / Formation of Surface texture	Wash #2
Diffusion of pn junction formation	Formation of p-type CIGS-based absorber
Isolation between the front and the back of the cell (Plasma etching)	Deposition of Buffer layer
AR Coating + Firing	Patterning #2
Front grid (Screen printing + Firing)	Deposition of n-type Window
Back electrode (Screen printing + Firing)	Patterning #3
	Finishing

I-V measurement

Module Processing

Strings for the interconnection

Packaging (Lamination)

I-V measurement

Fig. 11.15. Comparision of fabrication process between CIGS-based and crystalline-Si solar modules

Sekiyu K.K.) have demonstrated the efficiency of over 12% in a substrate size of 30 cm × 30 cm or larger [20–25], as shown in Table 11.5. This is good evidence to indicate the competitiveness of CIGS-based modules on the performance, because the efficiency of 12% is required as the lower limit of the distribution on the efficiency in the mass production stage. Therefore, CIGS-based modules are strongly required to demonstrate their superiority on the cost competitiveness by increasing a production volume in the module efficiency range of 12 to 13%.

At this moment, commercially available CIGS-based modules are ST-series of Shell Solar Industries [26], which are shown in Fig. 11.14, and WS-series of Würth Solar GmbH [25]. The status of the above three groups is as follows; 1) Shell Solar Industries is under expansion of their pilot production line for ST series modules. 2) ZSW GmbH has completed the technology transfer to Würth Solar GmbH and continuously functions as the technology development and technical support to Würth Solar GmbH [21, 25], which is in a pilot production stage and has a plan to expand the production volume of over 10 MW/y. 3) Showa Shell Sekiyu K.K. is still in the R&D stage focusing on development of "high-efficiency and low-cost" fabrication technologies applicable to the mass production.

Comparing the fabrication processes between CIGS-based and crystalline Si solar modules, it is noticeable that CIGS-based module technology has fewer steps as shown in Fig. 11.15. This is also well understood as an advantage of thin film technologies on the cost reduction. Thickness of a CIGS-based circuit with three patterns to make a monolithic structure is, in general, 3 to 4 μm excluding a substrate. This is about 100 times thinner than the thickness of crystalline Si solar cells, which are currently in a range of 300 to 400 μm (or 0.3 to 0.4 mm), although development of thinner crystalline Si solar cells is aggressively in progress. CIGS-based modules are, in general, fabricated by packaging or moduling a CIGS-based circuit or submodule, which is a stacked and monolithic structure as shown in Fig. 11.16, and by applying the packaging technologies for crystalline Si solar cells as shown in Fig. 11.17.

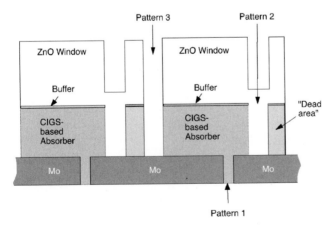

Fig. 11.16. Structure of monolithically integrated CIGS-based circuit

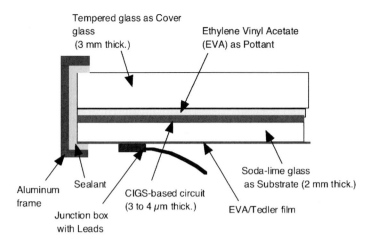

Fig. 11.17. CIGS-based module as a substrate structure by applying the packaging technologies for crystalline-Si solar cells

The energy payback time (EPT) calculated to the products in Siemens Solar Industries [27] also demonstrated the following advantages on the CIGS-based module technology; 1) the EPT was 1.8 years for CIGS-based module

Table 11.6. Fabrication technologies for a monolithically integrated CIGS-based circuit

Device structure	Shell Solar Ind.	ZSW GmbH/ Würth Solar GmbH	Showa Shell Sekiyu K.K.
Substrate	Substrate Preparation (Soda-lime glass)		
Base electrode	**Sputtering** (Mo)		
Pattern 1	**Laser**		
Absorber	**Sputtering** (Cu-Ga alloy/In as a stacked precursor) + Selenization /Sulfurization *(CIGSS)*	**Coevaporation** (One step method) (Cu, In, Ga, Se) *(CIGS)*	**Sputtering** (Cu-Ga alloy/In as a stacked precursor) + Selenization /Sulfurization *(CIGSS/CIGS)*
Buffer	**CBD** (CdS)		**CBD** (Zn(O,S,OH)$_x$)
Pattern 2	**Mechanical Scribing**		
Window (TCO)	**MOCVD** (ZnO:B)	**Sputtering** (ZnO:Al with thin intrinsic ZnO)	**Sputtering** (ZnO:Ga with thin intrinsic ZnO) or **MOCVD** (ZnO:B)
Pattern 3	**Mechanical Scribing**		

(Note) **CIGSS**= Cu(InGa)(SSe)$_2$, **CIGS**= Cu(InGa)Se$_2$, **CBD**= Chemical Bath Deposition, **TCO**= Transparent Conductive Oxide, **MOCVD**= Metal-Organic Chemical-Vapor Deposition.

(ST 40), while 3.3 years for single crystal (sc) Si module (SP 75). 2) Accumulated output power from the CIGS-based module for 30 years as an estimated lifetime was 17 times more than the consumed energy for its production, while 9 times more in the case of sc-Si solar module.

11.2.2 Fabrication Technologies

Monolithically integrated CIGS-based circuits as shown in Fig. 11.16, are fabricated with various technologies [20–25]. As shown in Table 11.6, similar device structure and common fabrication technologies are employed among three groups as a result of low-cost and large-area approaches. Among them, 1) different fabrication technologies are employed for p-type CIGS-based absorbers and n-type ZnO windows, while 2) different materials are deposited as a high-resistivity buffer. The fabrication technologies for both CIGS-based absorber and ZnO window [20–25, 28–33] still need some development to realize the mass production from the standpoint of continuous vs. batch process.

11.2.2.1 p-type CIGS-based Absorber

There are two formation processes for a large-area CIGS-based absorber as shown in Table 11.7. Limited by the transition point of soda-lime glass (i.e. 520–540°C dependent upon its thickness) as a substrate, the absorber formation temperature is carefully determined to avoid unfavorable deformation. Employing a one-step coevaporation method, the deformation of the substrate was prevented and the tact time was improved making the hold time at the highest temperature significantly shorter [21, 25]. ZSW GmbH successfully transferred this fabrication process to Würth Solar GmbH, which can make large-area absorbers up to 60 cm × 120 cm [25]. In the selenization/sulfurization method employed by Shell Solar Industries [20, 24, 34] and Showa Shell Sekiyu K.K. [22, 23, 35], the hardware dimension (i.e., a reaction furnace) affects the production volume because of a batch process.

11.2.2.2 High-resistivity Buffer

At this moment, most of high-performance CIGS-based solar cells and modules employ a CBD-CdS buffer as shown in Table 11.5. However, the approaches to make an eco-friendly device structure through the elimination of the toxic material like Cd are recognized as one of the important R&D subjects. The achievement by Showa Shell Sekiyu K.K. with a $Zn(O,S,OH)_x$ buffer indicates that high performance can be achieved without a CBD-CdS buffer. Many research groups demonstrate the high performance with various Cd-free buffers, described in detail in Sect. 11.1.3 [12, 22, 36–40]. Moreover, further disadvantage of using a CBD-CdS buffer is to need 1) a careful treatment of Cd-containing waste solution through the filtration [41] to separate a

Table 11.7. Formation processes for a large-area CIGS-based absorber

Method	Coevaporation (One step coevaporation)	Selenization/Sulfurization
Supply of Se	Se vapor (Continuous supply from a line source) [21,25]	1) Selenization of a sputtered Cu-Ga alloy/In stacked precursor layer with very low concentration of H_2Se gas, then sulfurization of CIGS with low concentration of H_2S gas [20,22,23,35]. 2) Annealing of an evaporated-Se/sputtered-Cu-Ga alloy/In stacked precursor layer with N_2 gas, then sulfurization of CIGS with low concentration of H_2S gas or Annealing of an evaporated-Se/sputtered-Cu-Ga alloy/In stacked precursor layer with low concentration of H_2S gas (Rapid Thermal Processing (RTP) method) [24,34].
Organization	ZSW GmbH and Würth Solar GmbH	1) Shell Solar Industries and Showa Shell Sekiyu K.K. 2) Shell Solar GmbH
Process	Continuous process in vacuum	Batch process in nearly atmospheric pressure
Maximum size of substrate	60cmx120cm	1) 30cmx120cm 2) 60cmx90cm
Remaining subjects	-to shorten the tact time. -to maintain the temperature uniformity.	-to develop a large-volume reactor. -to maintain the temperature uniformity.

liquid waste (neutral water) and a solid waste (CdS) and 2) an extra work to eliminate or wipe off the CdS from the rear side of the circuit from the safety point of view.

An interesting feature is a post-deposition light-soaking effect, which improves the device performance and can control in a reversible or irreversible form, adjusting the light soaking conditions [36, 37, 42].

11.2.2.3 n-type Window

As shown in Table 11.6, common deposition techniques for doped-ZnO TCO films as the n-type window are 1) sputtering with a ZnO:Al (2 wt % Al_2O_3) [33] or ZnO:Ga (5.7 wt % Ga_2O_3) ceramic target [29–32] and 2) MOCVD to prepare a ZnO:B film formed on a heated substrate at 140 to 200° through the reaction of diethyl zinc (DEZ) and deionized (DI) water with a very much low concentration of B_2H_6 gas as a dopant [28]. At this moment, sputtering as a kind of continuous process is believed to be more flexible to the mass production and much easier to pursue the low cost approaches, such as a reactive sputtering of Zn-Al alloy in $Ar + O_2$ mixed gas to make a ZnO:Al film [25, 33]. However, the electrical properties of doped-ZnO TCO films are remarkably different dependent upon the type of dopant. The mobility of sputtered-ZnO (i.e. ZnO:Al or ZnO:Ga) is almost half compared to that of MOCVD-ZnO (ZnO:B) [31]. In the monolithically-integrated CIGS-based circuits, sheet resistance of n-type window is critical and, in general, optimized about $10\Omega/\square$ as a function of transmittance (%T), or haze, because the photo-generated current in the CIGS-based absorber by incident light flows horizontally through the n-type window as seen in Fig. 11.16, in which each cell width is 5,000 to 15,000 times larger than the thickness of ZnO window.

11.2.2.4 Patterning

To prepare a monolithically-integrated CIGS-based circuit which consists of series-connected solar cells as shown in Fig. 11.16, the patterning techniques should be as follows: 1) to match the mechanical properties of each thin film in the CIGS-based circuit, 2) to be applied without any damage to the surrounding or adjacent area of patterns, and 3) to be performed to minimize the dead area (or loss of an effective region) of each CIGS-based solar cell. As shown in Table 11.6, two types of patterning techniques with a laser and a mechanical scriber are currently applied to make an integrated circuit with three patterns from the standpoint of low cost, high speed and reliability. The requirements to the patterning are summarized in Table 11.8.

11.2.2.5 Process Control

Development of process control techniques and standardized procedure is one of the key issues to realize the mass production and to improve the reproducibility and robustness in the fabrication process for CIGS-based modules.

Table 11.8. Requirements to the patterning

Patterning	Components to pattern	Requirement
Pattern 1 = Laser	Base electrode (Mo)	-to develop the laser process without making any shunts. -to develop the Mo suitable to laser processing. -to keep good adhesion at the interface of Mo/glass substrate and Mo/absorber.
Pattern 2 = Mechanical scriber	Buffer/CIGS-based absorber	-to develop a process suitable to the soft material like absorber. -to develop a low-cost and high-speed processing.
Pattern 3 = Mechanical scriber	Window	-to develop a crack-free process applicable to the hard material like ZnO. -to develop a low-cost and high-speed processing.

Table 11.9. Process control parameters in the fabrication of CIGS-based circuits

Device Structure	Organization		
	Shell Solar Industries	ZSW GmbH	Showa Shell Sekiyu K.K.
	Process Control Parameters		
Soda-lime glass substrate	Check of front or back		
Mo Base electrode	Sheet resistance and Thickness	Sheet resistance, Thickness and Adhesion	Sheet resistance, Thickness and Reflectance
Pattern 1	-	Isolation	
p-type CIGS-based Absorber	Cu/III Ratio (Precursor layer)	Cu/III Ratio, Sheet resistance and Thickness (Absorber)	Cu/III Ratio and Thickness (Precursor layer), S/VI Ratio and Reflectance (Absorber)
CBD Buffer	-	[Visual inspection]	Transparency (or Transmittance) of CBD solution
Sputtered high-resistivity ZnO (i-ZnO)	-	Sheet resistance, Thickness and Transmittance	
Pattern 2	-	Isolation	[Sampling to inspect with optical microscope]
n-type ZnO Window (TCO)	Sheet resistance and Thickness	Sheet resistance, Thickness and Transmittance	Sheet resistance, Thickness and Transmittance (if necessary, Carrier conc. and Mobility)
Pattern 3	-	Isolation	[Sampling to inspect with optical microscope]

The parameters for process control are shown in Table 11.9. The concepts on the process control should be different between R&D and production stages, because a production stage needs much quicker feedback and smaller number of control parameters to maintain the high throughput. One approach that Shell Solar Industries has employed is the statistical process control (SPC) [43, 44] and other groups have also tried the same procedure in a much smaller production volume [21, 45, 46], because the mechanical and electrical yields are a good yardstick to evaluate the robustness and reproducibility of the fabrication process. A required yield should be 85% even in the initial stage of production. By applying the slicing to prepare small size of circuits, it is also effective to improve the yield and reduce the production cost. For instance, the circuits of 30 cm × 15 cm (ST5), 30 cm × 30 cm (ST10) and 30 cm × 60 cm

(ST20) are fabricated by slicing a 30 cm × 120 cm-sized circuit (ST40) in the case of Shell Solar Industries [20, 34], which is a significant advantage of thin film technologies.

11.2.3 Durability

11.2.3.1 Outdoor Exposure Test

At an initial stage, the packaging techniques good for crystalline Si solar cells were applied to CIGS-based circuits and some troubles (for instance, temporary degradation by moduling and recovery by light soaking) had been reported [47]. Therefore, adjustment of the packaging technologies for crystalline Si solar cells was carried out to prepare stable CIGS-based modules. The 1 kW system of ST40 (30 cm × 120 cm)-sized CIGS-based modules, which was prepared by Shell Solar Industries (formerly Siemens Solar Industries) and installed at the outdoor exposure test site in the NREL, has demonstrated stable performance for last 10 years [20]. Based upon this result, Shell Solar Industries has set the warrantee of ST series modules for 10 years [26] and this duration will be longer. Furthermore, Shell Solar Industries has accumulated the outdoor exposure test data of the average 1 kW system from the 9 selected sites of 8 in the USA including the NREL site and 1 in Germany, based upon the different climate conditions, under the financial support by the California Energy Commission (CEC). At this moment, the largest CIGS-based module system, which was installed by Shell Solar GmbH in the newly constructed Congress Hall in Salzburg, Austria, is a 40.8 kW system with 1020 (i.e. 960 on the roof and 60 on the façade) ST40 modules by Shell Solar Industries and expected to supply about 35,000 kWh/y [26]. Such large installations are currently seen in Europe, for instance Switzerland and Germany and will be increasing more and more in the near future by decreasing the module cost. To accelerate the market penetration and to fulfill further cost down requirement, it is important to develop alternative low-cost and high-reliability packaging technologies suitable to CIGS-based circuits, because the moduling cost in the case of applying the fabrication techniques of crystalline Si solar modules directly to CIGS-based circuits is estimated as a major part of the production cost of CIGS-based module.

11.2.3.2 Safety Issue

It is well known that CIGS-based absorber is insoluble in acidic and alkaline solutions. In the acid-rain condition (pH $= 4$), leaching experiments from smashed CIGS-based modules were confirmed not to detect unfavorable metal ions from a CIGS-based absorber over the threshold limit [48]. In the alkaline solution, CIGS-based absorber is immersed in a hot strong-alkaline (pH ≥ 10) solution to deposit a high-resistivity buffer by the CBD techniques. This is a good evidence that CIGS-based absorber is insoluble in alkaline solution. Combustion experiments on the possible fire to the CIGS-based module systems and animal tests on an intake amount of CIGS-based solar cells in a human body were performed in the Brookhaven National Lab. [49–52], European and German Programs [53] and New Sunshine Program in Japan [48]. All reports concluded that there were no anxieties on the safety issues with CIGS-based modules.

11.2.3.3 Space Application

Space is a new frontier to CIGS-based modules. Requirements to the solar cells in the space application are as follows;

1. high radiation hardness on electron and proton,
2. high output per weight (W$_p$/kg) (or high output + light weight),
3. high potential to reduce the cost.

In 1997, the EQUATOR-S scientific satellite was launched, which was a spin-stabilized satellite to study the earth's equatorial magnetosphere on an elliptical orbit (500/65600 km) for two years, and it was confirmed from the flight experiments operated in a proton-dominated orbit that CIGS-based mini-modules with 6 monolithically integrated cells on a 4 cm × 4 cm glass substrate had significantly high radiation hardness on electron and proton and showed an improved performance compared to the ground data [54]. Based upon these results, the high radiation hardness on electron and proton of CIGS-based absorber was concluded to originate from the ability to accommodate the defects formed by the impact of electron and proton, in which the defects formed electrically neutral complexes (for instance, $2V_{Cu}^- - In_{Cu}^{2+}$) with no contribution to the electrical performance [55–57]. The National Aero Space Development Association in Japan (NASDA), obtained the same trends through the irradiation experiments, in which CIGS-based solar cells demonstrated no degradation against the electron and smaller degradation against the proton than other solar cells [58]. This was a noteworthy evidence that CIGS-based solar cells and modules were the best candidate in the space as a power source. To prepare the era frequently launching the low-cost satellites which were assembled with low cost parts to reduce the satellite cost, NASDA has a plan to collect the irradiation data of various terrestrial solar cells, which were mounted on the TSC-S of the MDS-1 satellite, named as Tsubasa after launch on February 2002, by operating in a geostationary transfer orbit (500/36000 km) [59].

CIGS-based solar cells and modules as a thin-film solar cell have the highest potential among the existing solar cells to reduce the cost or improve the power per weight (W$_p$/kg), because CIGS-based solar cells and modules demonstrate 1) the highest performance among the thin-film solar cells [60–62] and 2) a high potential to reduce the weight remarkably by applying various light weight substrates, such as stainless steel foil or plastic [63–66]. Currently, the R&D on CIGS-based devices is focused on understanding why CIGS-based devices demonstrate higher radiation hardness on electron and proton than other space solar cells and how these superior properties of CIGS-based devices is further improved.

References

1. Mickelsen, R.A., Chen, W.S., Hsiao, Y.R., Lowe, V.E. (1984): IEEE Trans. Electron Devices, 31, 542.
2. Contreras, M.A., Egaas, B., Ramanathan, K., Hiltner, J., Swartzlander, A., Hasoon, F., Noufi, R. (1999): Prog. in Photovoltaics, 7, 311.
3. Kushiya, K., Tachiyuki, M., Nagoya, Y., Fujimaki, A., Sang, B., D. Okumura, D., Satoh, M., Yamase, O. (2001): *Solar Energy Mater. Solar Cells*, 67, 11.
4. Wada, T. (1997): *Solar Energy Mater. Solar Cells*, 49, 249.
5. Nishiwaki, S., Kohara, N., Negami, T., Wada, T.: (1998) *Jpn. J. Appl. Phys.*, 37, L71.
6. Kohara, N., Nishitani, M., Negami, T., Terauchi, M., Ikeda, M., Wada, T. (1995): Proc. 13th EC-PVSEC, 23 Oct., Nice p.2084.
7. Kohara, N., Negami, T., Nishitani, M., Wada, T. (1995): *Jpn. J. Appl. Phys.*, 34, L1141.
8. Kitagawa, K., Negami, T. (2001): Tech. Digest of 14th Sunshine Workshop, 1, Feb.,Tokyo p.86.
9. Ramanathan, K., Wiesner, H., Asher, S., Niles, D., Bhattacharya, R.N., Keane, J., Contreras, M.A., Noufi, R. (1998): Proc. 2nd WCPEC, 6, July, Vienna p.477.
10. Wada, T., Hayashi, S., Hashimoto, Y., Nishiwaki, S., Sato, T., Negami, T., and Nishitani, M. (1998): Proc. 2nd WCPEC, 6, July, Vienna p.403.
11. Nakada, T., and Mizutani, M. (2000): Proc. 28th IEEE PVSC, Anchorage, 15–22 Sept. p.529.
12. Ohtake, Y., Chaisitsak, S., Yamada, A., Konagai, M. (1998): *Jpn. J. Appl. Phys.*, 37, 3220.
13. Chaisitsak, S., Yamada, A., Konagai, M., Saito, K. (2000): *Jpn. J. Appl. Phys.*, 39,1660.
14. Negami, T., Minumoto, T., Hashimoto, Y., Satoh, T. (2000): Proc. 28th IEEE PVSC, Anchorage, 15-22 Sept. p.634.
15. Satoh, T., Hashimoto, Y., Shimakawa, S., Hayashi, S., Negami, T. (2001): Tech. Digest of PVSEC-12, Jeju, Korea p.93.
16. Hartmann, M., Schmidt, M., Jasenek, A., and Schock, H.-W. (2000): Proc. 28th IEEE PVSC, Anchorage, 15–22 Sept. p.638.
17. Tiwari, A.N., Krejci, M., Haug, F.J., Zogg, H. (1999): *Prog. Photovolt.*, 7, 393.
18. Bailey, S., and Hepp, A. (2001): Tech. Digest of PVSEC-12, Jeju, Korea p.571.
19. Tuttle, J.R., Szalaj, A., Keane, J. (2000): Proc. 28th IEEE PVSC, Anchorage, 15–22 Sept. p.1042.
20. Tarrant, D.E., Wieting, R.D. (2001): Tech. Digest 14th "Sunshine Workshop" on Thin Film Solar Cells p.80.
21. Powalla, M., and Dimmler, B. (2003): *Sol. Energy Mater. Sol. Cells* 75 p.27.
22. Kushiya, K., Ohshita, M., Hara, I., Tanaka, Y., Sang, B., Nagoya, Y., Tachiyuki, M., Yamase, O. (2003): *Sol. Energy Mater. Sol. Cells* 75 p.171.

23. Tanaka, Y., Akema, N., Morishita, T., Okumura, D., Kushiya, K. (2001): Proc. 17th ECPVSEC p.989.
24. Probst, V., Stetter, W., Riedl, W., Vogt, H., Wendl, M., Calwer, H., Zweigart, S., Ufert, K.-D., Freienstein, B., Cerva, H., Karg, F.H. (2002): Proc. Symposium **N** on Thin Film Chalcogenide Photovoltaic Materials Vol.**387** p.262.
25. Powalla, M., Dimmler B. (2001): Proc. 17th ECPVSEC p.983.
26. Shell Solar Industries, ST series catalogs.
27. Knapp, K.E., Jester, T.L. (2000): Proc. 28th IEEE PVSC p.1450.
28. Pier, D., Gay, C.F., Wieting, R.D., Langeberg, H.J.: US Patent No. 5,078,803.
29. Cooray, F.W., Kushiya, K., Fujimaki, A., Okumura, D., Sato, M., Ooshita, M., Yamase, O. (1999): *Jpn. J. Appl. Phys.* **38** p.6213.
30. Kushiya, K., Sang, B., Okumura, D., Yamase, O. (1999): *Jpn. J. Appl. Phys.* **38** p.3997.
31. Sang, B., Kushiya, K., Okumura, D., Yamase, O. (2001): *Sol. Energy Mater. Sol. Cells* **67** p.237.
32. Nagoya, Y., Sang, B., Fujiwara, Y., Kushiya, K., Yamase, O. (2003): *Sol. Energy Mater. Sol. Cells* **75** p.163.
33. Ruckh, M., Hariskos, D., Rühle, U., Schock, H.W., Menner, R., Dimmler, B. (1996): Proc. 25th IEEE PVSC p.825.
34. Karg, F. (2001): *Sol. Energy Mater. Sol. Cells* **66** p.645.
35. Kushiya, K., Kase, T., Tachiyuki, M., Sugiyama, I., Satoh, Y., Satoh, M., Takeshita, H. (1996): Proc. 25th IEEE PVSC p.989.
36. Kushiya, K., Nii, T., Sugiyama, I., Sato, Y., Inamori, Y., Takeshita, H. (1996): *Jpn. J. Appl. Phys.* **35** p.4383.
37. Kushiya, K., Tachiyuki, M., Kase, T., Sugiyama, I., Nagoya, Y., Okumura, D., Sato, M., Yamase, O., Takeshita, H. (1997): *Sol. Energy Mater. Sol. Cells* **49** p.277.
38. Hariskos, D., Ruckh, M., Rühle, U., Walter, T., Schock, H.W., Hedström, J., Stolt, L. (1996): *Sol. Energy Mater. Sol. Cells* **41/42** p.345.
39. Eisele, W., Ennaoui, A., Bischoff, P.S., Giersig, M., Pettenkofer, C., Krauser, J., Lux-Steiner, M., Riedle, Esser, T., N., Zweigart, S., Karg, F. (2000): Proc, 28th IEEE PVSC p.692.
40. Rumberg, A., Gerhard, A., Jäger-Waldau, A., Lux-Steiner, M. Ch. (2003): *Sol. Energy Mater. Sol. Cells* **75** p.1.
41. Hariskos, D., Powalla, M., Chevaldonnet, N., Lincot, D., Schindler, A., Dimmler, B. (2000): Proc. Symposium **N** on Thin Film Chalcogenide Photovoltaic Materials Vol.**387** p.179.
42. Kushiya, K., Yamase, O. (2000): *Jpn. J. Appl. Phys.* **39** p.2577.
43. Wieting, R., DeLaney, D., Dietrich, M., Fredric, C., Jensen, C., Willett, D. (1995): Proc. 13th ECPVSEC p.1627.
44. Ulfert, J., Wieting, R. (2000): Proc. 28th IEEEPVSC p.466.
45. Powalla, M., Dimmler, B. (2000): Proc. Symposium **N** on Thin Film Chalcogenide Photovoltaic Materials Vol.**387** p.251.
46. Kushiya, K., Hara, I., Tanaka, Y., Morishita, T., Okumura, D., Nagoya, Y., Tachiyuki, M., Sang, B., Yamase, O. (2000): Proc. 28th IEEEPVSC p.424.
47. Tarrant, D. (1991): *Solar Cells* **30** p.549.
48. Central Research Institute of Electric Power Industry (CRIEPI) (2001), FY2000 Annual Report of NEDO New Sunshine program. Survey study on reliability improvement of photovoltaic power generation (Survey study on environmental impact issues of compound solar cells and modules) p.39-91 (in Japanese). (NEDO Home page, http://www.tech.nedo.go.jp/)

49. Moskowitz, P.D., Ftehnakis, V.M. (1990): *Solar Cells* **29** p.63.
50. Thumm, W., Finke, A., Neumeier, B., Beck, B., Kettup, A., Steinberger, H., Moskowitz, P.D. (1994): Proc. 1st WCPEC p.262.
51. Ftehnakis, V.M., Morris, S.C., Moskowitz, P.D., Morgan, D.L. (1999): *Prog. Photovolt: Res. Appl.* **7** p.489.
52. Moskowitz, P.D., Ftehnakis, V.M. (1991): *Solar Cells* **30** p.89.
53. Powalla, M., Dimmler, B. (2000): *Thin Solid Films* **361-362** p.540.
54. Schock, H.W., Bogus, K. (1998): Proc. 2nd WCPEC p.3586.
55. Jasenek, A., Rau, U. (2001): *J. Appl. Phys.* **90** p.650.
56. Jasenek, A., Schock, H.W., Werner, J. H., Rau, U. (2001): *Appl. Phys. Lett* **79** p.2922.
57. Guillemoles, J.-F., Rau, U., Kronik, L., Schock, H.W., Cahen, D. (2000): *Adv. Mat.* **8** p.111.
58. Hisamatsu, T., Aburaya, T., Matsuda, S. (1998): Proc. 2nd WCPEC p.3568.
59. NASDA, MDS-1 Home page,
    `http://www.nasda.go.jp/projects/sat/mds/tback_j.html`
60. Contreras, M., Egaas, B., Ramanathan, K., Hiltner, J., Swartzlander, A., Hasoon, F., Noufi, R. (1999): *Prog. Photovolt: Res. Appl.* **7** p.311.
61. Negami, T., Satoh, T., Hashimoto, Y., Nishiwaki, S., Shimakawa, S., Hayashi, S. (2001): *Sol. Energy Mater. Sol. Cells* **67** p.1.
62. Hagiwara, Y., Nakada, T., Kunioka, A. (2001): *Sol. Energy Mater. Sol. Cells* **67** p.267.
63. Tuttle, J.R., Szalaj, A., Keane, J. (2000): Proc. 28th IEEE PVSC p.1042.
64. Ullal, H.S., Zweibel, K., von Roedern, B.G. (2000): Proc. 28th IEEE PVSC p.418.
65. Hartmann, M., Schmidt, M., Jasenek, A., Schock, H.W., Kessler, F., Herz, K., Powalla, M. (2000): Proc. 28th IEEE PVSC p.638.
66. Satoh, T., Hashimoto, Y., Shimakawa, S., Hayashi, S., Negami, T. (2003): *Sol. Energy Mater. Sol. Cells* **75** p.65.

12

Expanding Thin-Film Solar PV System Applications

Hirosato Yagi, Makoto Tanaka, and Shoichi Nakano

Applications of thin-film solar cells, especially a-Si, started in the 1980s mainly in the field of electrical devices. By utilizing the features of thin-film solar cells, unique applications such as larger electrical power sources have been developed. Three kinds of applications are introduced in this chapter. The first one is the see-through a-Si solar cell, the second is the flexible a-Si solar cell, and the final one is the building integrated PV (BIPV) module. In the future, thin-film solar cells are expected to play an important role for the widespread use of solar cells from the field of electrical devices to the field of electrical generating use on a global scale.

12.1 Introduction

In recent times, the market for photovoltaic power-generation systems has grown rapidly. The production volume of solar cells in 2000 was more than 280 MW/y, which was 40% larger than that of the previous year. Especially in Japan, production volume has increased by 50%, to 120 MW. This is mainly because of the subsidies provided by the Japanese government for photovoltaic systems.

The majority of the solar cells used in photovoltaic system are crystalline-silicon solar cells. The main application of thin-film solar cells had been consumer products such as pocket calculators; however, their use in power generation has recently increased.

The history of thin-film solar cells is different from that of crystalline-silicon solar cells with respect to their applications. As thin-film solar cells were industrialized after crystalline-silicon cells, they should be positioned in a more specialized market in which features such as their high voltage, light weight, etc., are utilized.

In this chapter, a comparison of the market between thin-film solar cells and crystalline-silicon solar cells is given first. Next, various kinds of applications of thin-film solar cells are presented. Finally, future prospects are discussed.

12.2 Comparison Between Crystalline Solar Cells and Thin-Film Solar Cells with Respect to Applications

A comparison of applications between crystalline-silicon solar cells and thin-film solar cells is shown in Fig. 12.1.

The first applications of crystalline-silicon solar cells were in space use, telecommunication use, etc., which were stand alone systems. The grid-connected market started in Japan in 1992. Since then, most power-generation systems have consisted of normal types of modules. Special modules, such as BIPV (building integrated PV) modules, semitransparent modules, etc., were of minor significance although they have been increasing recently.

On the other hand, thin-film solar cells were industrialized in the special markets of consumer electronics, see-through solar cells for windows applications, etc. In these applications, the features of thin-film solar cells are utilized (see Sect. 12.3). Figure 12.2 shows the features and the applications of thin-film solar cells. The first market for thin-film solar cells was pocket calculators. Here, the following features of thin-film solar cells were utilized; high voltage can be obtained easily from a single substrate and high conversion efficiency is obtained in fluorescent light.

See-through solar cells and BIPV modules were the main applications in power-generation use, where the features of thin-film solar cells with good appearance and a simple module structure, etc., are utilized.

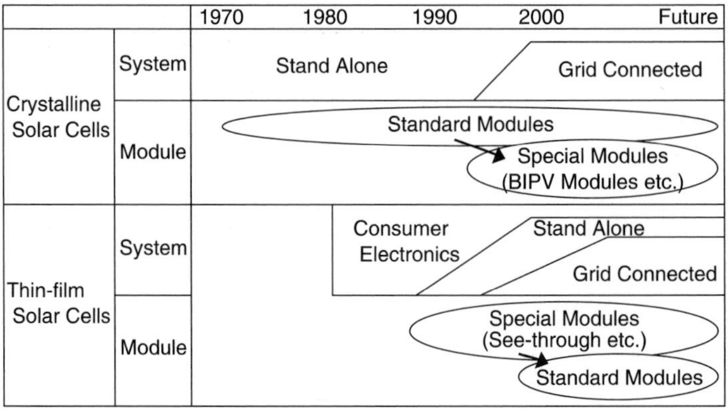

Fig. 12.1. Applications of crystalline solar cells and thin-film solar cells

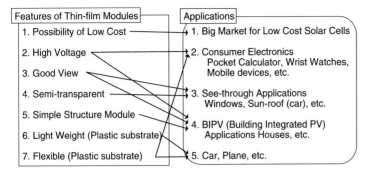

Fig. 12.2. Features of applications of thin-film solar cells

Flexible thin-film modules have also been industrialized. These modules have the features of flexibility and light weight. At present, their market is relatively small, however, their market is expected to grow in the future.

The most prominent feature, low cost, will create a substantial market for power generation in the future.

12.3 Applications to Electrical Devices

Since the 1980s, solar cells, especially a-Si solar cells, have enjoyed widespread use as a power source mainly for electrical devices, such as pocket calculators, watches, etc. (Fig. 12.3). There are three reasons for the widespread use of a-Si solar cells in the area of electrical devices.

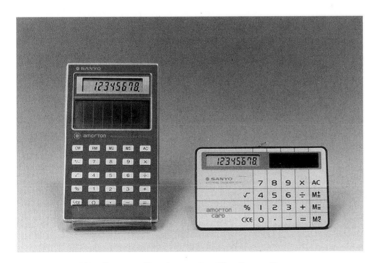

Fig. 12.3. The first applications of a-Si solar cell

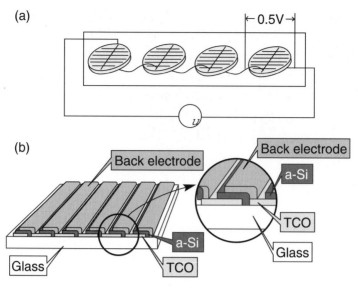

Fig. 12.4. (a) Conventional-type solar cell module and (b) integrated-type a-Si solar cell modules

The first reason comes from the excellent feature of a-Si solar cells as thin-film solar cells. a-Si solar cells can generate high-voltage electrical power with an integrated-type structure as shown in Fig. 12.4 [1]. Three layers of a-Si solar cells, the TCO (transparent conductive oxide) layer, the a-Si layer and a metal layer, are separated in the fabrication process so that neighboring separated cells are connected consecutively at the edges of the metal layer and TCO layer. Where c-Si solar cells are used, each cell should be connected consecutively by soldering.

The second reason is that a-Si solar cells are conducive to indoor lighting. In Japan, fluorescent lamps are commonly used for indoor lighting. Fluorescent light has a high intensity in the short-wavelength region and a-Si solar cells have high sensitivity in this region. This means that a-Si solar cells are more suitable for indoor use compared with c-Si solar cells.

The third reason is dependent on the electrical device itself. Through the progress of semiconductor technology, the size of transistors in IC and LSI has greatly reduced and the electrical power required for operation has diminished. This has accelerated the application of a-Si solar cells for indoor use.

Figure 12.5 shows the applications of a-Si solar cells to electrical devices, pocket calculators, watches, clocks, radios, testers, TV controllers, battery chargers, etc.

In particular, a-Si solar cells occupy a very large part of the market in the application to pocket calculators. Recently, an application to watches has become very promising.

Fig. 12.5. Applications of a-Si solar cells

12.4 Applications to Standalone Systems

Standalone systems with a capacity ranging from a few tens of kW to a few kW have been developed. In daytime, solar cells charge the battery and a rechargeable battery supplies electricity to the load at night. Figure 12.6 shows an example of a standalone system. On the right of Fig. 12.6, a "solar sensor light", a lighting system composed of an IR sensor, a-Si solar cells, and battery, is presented. To the left of that, a "solar planter", a hydroponics home farming kit in which solar cells are used to drive an air pump, is indicated. Figure 12.7 shows an application called the "solar guide post", which incorporates a-Si solar cells and a rechargeable battery on the top of a roadside mailbox. At

Fig. 12.6. Example of standalone system

Fig. 12.7. Solar guide post

night, electricity is supplied from the battery to light up the map and postal symbol mark.

12.5 See-Through a-Si Solar Cells

Special solar-cell modules were developed by utilizing the features of thin-film solar cells. Figure 12.8 shows three kinds of these: see-through solar cells, flexible solar cells, and BIPV modules.

See-through a-Si solar cells are semitransparent solar cells. They can generate electricity while transmitting part of the incidental light to the back

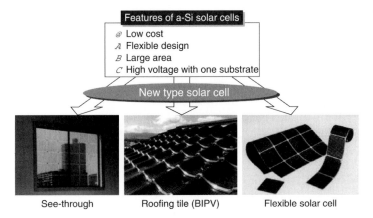

Fig. 12.8. Unique applications of a-Si solar cells

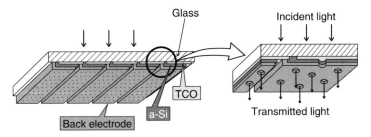

Fig. 12.9. Structure of see-through a-Si solar cell

side [2]. Figure 12.9 shows the structure of a see-through a-Si solar cell. One salient feature of the structure is the uniform presence of multiple microscopic holes in the effective area of integrated-type a-Si solar cells. These holes are formed by removing the a-Si layer and metal layer by chemical etching or by a laser. Part of the incidental light can pass through these holes.

There is another type of semitransparent a-Si solar cell in which the metal layer is replaced by the TCO layer (Fig. 12.10b). But in this type of a-Si solar cell the transmitted light becomes reddish in color because of the absorption of short-wavelength light (500–650 nm) by the a-Si layer, as shown in Fig. 12.10c. On the other hand, the spectral transmittance of see-through a-Si solar cells remains almost constant over a wide wavelength region. Therefore, natural interior lighting can be obtained and the rate of transmittance can be controlled in a design value by changing the area of a hole.

Figure 12.11 shows an application to car sunroofs. Electrical power generated by solar cells is used to drive a ventilation fan. It is effective in reducing the temperature inside a parked car on a sunny summer day. Another promis-

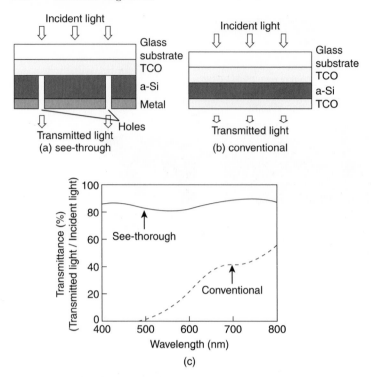

Fig. 12.10. (a-c). Comparison between see-through a-Si solar cell and conventional one

ing application of see-through a-Si solar cells is their use as the windows of buildings (see Fig. 12.20).

12.6 Flexible a-Si Solar Cells

Another new type of solar cell utilizing the features of a-Si solar cells is the flexible a-Si solar cell. Flexible a-Si solar cells are a-Si solar cells fabricated on a plastic film [3]. They are very light and flexible. Figure 12.12 shows the structure of a flexible a-Si solar-cell module. Figure 12.13 shows the appearance of a flexible a-Si module. The weight per square meter is about 500 g and the power-weight ratio is about 0.1 W g^{-1}. Compared with those of c-Si, the weight per square meter is about one twentieth and the power-weight ratio is about one tenth. Figure 12.14 shows an airplane powered by thin-film solar cells.

It seems that the most promising application of flexible a-Si solar cells is their use as a portable power source. For example, solar-powered portable phones, parasols, and material integrated with flexible a-Si solar-cell modules.

Fig. 12.11. See-through sunroof

12.7 Applications for Residential Housing (Building Integrated PV Modules)

12.7.1 Present Market for PV Housing

Residential housing is the main market in the world, especially in industrialized countries. In Japan and Germany, subsidies for photovoltaic systems for residential housing have created a huge market. The main modules used are crystalline-silicon solar cells, however, thin-film solar cells have begun to be employed (Fig. 12.15).

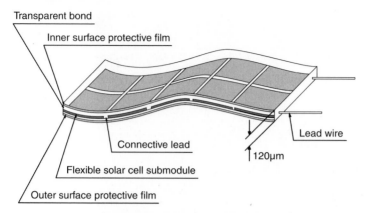

Transparent bond
Inner surface protective film
Lead wire
Connective lead
120µm
Flexible solar cell submodule
Outer surface protective film

Fig. 12.12. Structure of flexible a-Si solar-cell module

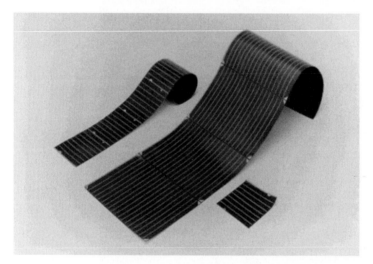

Fig. 12.13. Appearance of flexible a-Si solar-cell module

Recent features of residential PV systems can be found in the building integrated PV (BIPV) modules. In Japan, BIPV modules were certified as building materials in 1999 under article 38 of the Building Standard Law. Since then, BIPV modules have been widely used for residential housing.

12.7.2 Development BIPV Modules

In the early period, BIPV modules were made of thin-film solar cells because, as noted before, thin-film solar cells have features such as good appearance, simple structure suited for BIPV modules etc.

Fig. 12.14. Solar plane

Fig. 12.15. Example of residential house installed with a-Si solar-cell modules

In the mid 1980s, Japanese-style roofing tiles were developed. This module has the same shape as conventional Japanese roofing tiles (Fig. 12.16) with a curved surface, which were only realized by thin-film solar cells.

In the 1990s, many manufacturers of PV modules started R&D for BIPV modules. Most of those in Japan were supported by NEDO (New Energy and Industrial Technology Development Organization). Figure 12.17 shows an example of BIPV modules using amorphous-silicon solar cells. Recently, the stabilized conversion efficiencies of amorphous-silicon solar cells have been increased to 10%, and the conversion efficiencies of other thin-film solar cells

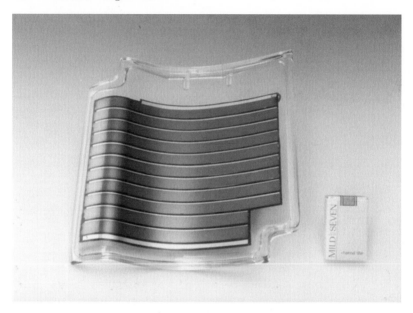

Fig. 12.16. a-Si solar tile

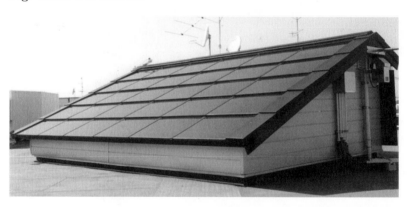

Fig. 12.17. Example of BIPV module (a-Si)

have also been improved. Therefore, BIPV thin-film modules will be utilized in various markets.

12.7.3 Industrialization of BIPV Modules

The industrialization of BIPV modules was started in thin-film solar cells. See-through solar cells were the first BIPV thin-film solar cells. In 1997, BIPV modules using amorphous-silicon solar cells were industrialized (see Chaps. 6 and 8). All of the products made by this type of manufacturing were BIPV modules.

Fig. 12.18. Example of BIPV modules (HIT)

New modules have also been industrialized for crystalline-silicon solar cells. Figure 12.18 shows an example of BIPV module applications using crystalline-silicon solar cells. Here, the solar cells are HIT (heterojunction with intrinsic thin layer) solar cells, which incorporates a-Si technology [4]. Figure 12.19 shows another example of BIPV module using amorphous-silicon solar cells.

12.8 Application to Semi-Large-Scale Photovoltaic Systems

In the field of semi-large-scale photovoltaic systems, c-Si solar cells are mainly used. However, a-Si solar cells are used in some cases in which the appearance of the photovoltaic system is important or solar cells are installed in the windows and/or skylight. Figure 12.20 shows a photovoltaic system in the Hamaoka Nuclear Exhibition Center of the Chubu Electric Power Co., Inc. See-through solar cell modules (1.95 kW) have been installed in the windows of the building. Figure 12.21 is another example from the Tukasa Electric Co., Ltd. of Osaka prefecture, Japan. See-through a-Si solar cell modules (1.4 kW)

Fig. 12.19. Example of BIPV modules (a-Si)

Fig. 12.20. Example of PV system (1) (see-through: Japan)

Fig. 12.21. Example of PV system (2) (see-through: Japan)

Fig. 12.22. Example of PV system (3) (CIS: Switzerland) (presented by FABRi SOLAR AG)

have been installed in windows and conventional type a-Si solar-cell modules (3.7 kW) have been installed in one of the building walls. Figure 12.22 shows a PV system in Switzerland with CIS (copper indium selenide) modules.

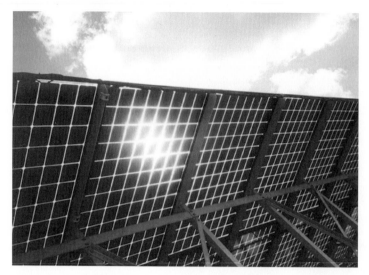

Fig. 12.23. Bifacial solar cell module with HIT solar cells

Fig. 12.24. Application to a portable phone

Unique structure modules are also applicable to semi-large-scale photo-voltaic systems. Figure 12.23 shows a bifacial solar-cell module with HIT solar cells, which can generate electricity on the back side as well as on the front side. This type of module is expected to be very effective when modules are installed in the vertical soundproof walls of highways or railways.

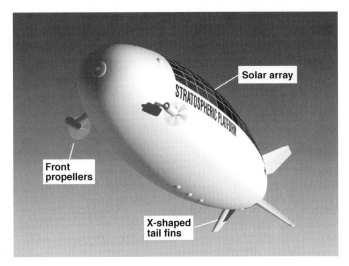

Fig. 12.25. Solar-powered airship in stratosphere [5]

12.9 Future Prospects

As noted previously, thin-film solar cells have many prominent features and various kinds of market are expected.

12.9.1 Applications of Flexible and Lightweight Modules

The first promising market is the application of lightweight and flexible solar cells. Recently, mobile digital equipment such as portable phones, PDAs (personal digital assistance), pocket game machines, etc., have been widely used all over the world. Lightweight solar cells are useful for these types of mobile equipment. Figure 12.24 shows one example, which is a portable phone. At present, there are only a few applications in this field because the generating power is insufficient (as these types of mobile equipment are so small, the size of the solar cells must also be small). However, the conversion efficiency of the solar cells will be improved and the consumption power of the mobile equipment will decrease.

Other applications of the lightweight modules are expected for automobiles, airships, and airplanes. Figure 12.25 shows a design concept of a stratospheric platform airship in which ultra-lightweight solar cells are employed to provide daytime/night propulsion power.

12.9.2 Applications of BIPV Modules

BIPV modules are also promising modules. In addition to rooftops and windows, solar-cell modules will be installed at other locations such as sidewalls. Figure 12.26 shows a future house powered by thin-film solar cells.

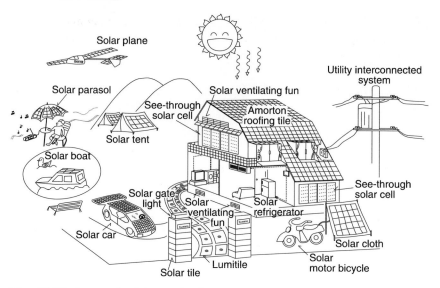

Fig. 12.26. Image of a future solar house

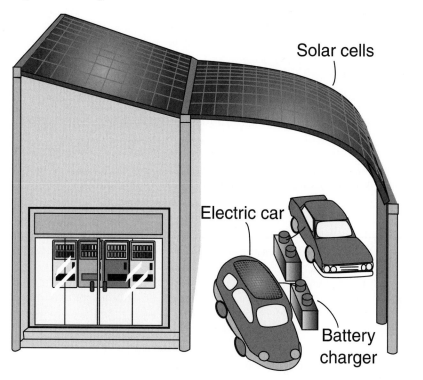

Fig. 12.27. Image of PV system for electric cars

For industrial applications, the soundproofing of highways and the side-walls of buildings are other candidates for solar-cell application. Figure 12.27 shows a future image.

12.9.3 Applications of Power-Generating Use – GENESIS Project

The most promising feature of thin-film solar cells is the low cost. At present, the market for solar cells is about 280 MW/y, which is much smaller than that of the total demand of electricity. For example, although Japanese peak demand is about 170 GW, the Japanese photovoltaics market is 120 MW. Of all the features, that of low-cost is the most important.

Described next is the GENESIS Project, which Sanyo proposes as a future energy-generation system based on solar power-generation [6].

The main disadvantages of solar cells are their inability to generate power at night, and that their output fluctuates dramatically depending on the sunlight conditions. These problems give some concern to those who feel that solar energy is unstable as a prime energy source.

As outlined in Fig. 12.28, a global energy network equipped with solar cells and international superconductor grids (GENESIS) has been proposed by the author for resolving these problems. A look at the Earth from outer space shows that rainy and cloudy areas cover less than 30% of the total land mass, and obviously it is always daylight on the opposite side of the globe to those areas under the shade of night. A worldwide photovoltaic power-generating system connected by super-conducting cables with no transmission losses would enable daylight areas to provide clean solar energy around the clock to those areas where it is night, rainy or cloudy. This would ensure that no area on Earth is without power.

It is forecast that in the year 2010, energy demands will be the equivalent of 14 billion kl of crude oil per year. To meet this requirement, 802 km^2 of solar cells will be required, assuming a conversion efficiency of 10%. This plan

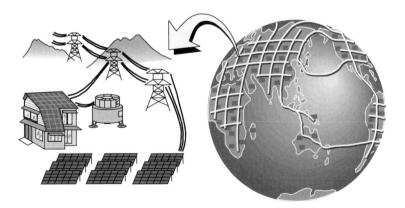

Fig. 12.28. GENESIS project

is quite feasible because barely 4% of the world's desert area would suffice. In the midst of the worsening energy crisis and environmental concerns, this plan must be put into effect for the sake of a prosperous 21st century.

Acknowledgment

This work is supported in part by the New Energy and Industrial Technology Development Organization (NEDO) as part of the New Sunshine Program under the Ministry of Economy, Trade, and Industry.

References

1. Kuwano, Y., et al. (1981): U.S Patent 4, 281, 208.
2. Yagi, H., et al. (1998): New module technologies for solar cells. The 2nd Korea-Japan Joint Seminar, pp.107–113.
3. Kishi, Y., et al. (1992): A new type of ultra light flexible a-Si solar cell. *Jpn. J. Appl. Phys.* 31, pp.12–17.
4. Kawamoto, K., et al. (2001): A high efficiency HIT solar cell (21.0%, 100 cm^2) with excellent interface properties. PVSEC 12, KOREA, pp.289–290.
5. Eguchi, K., et al. (2001): Overview of Stratospheric Platform Airship R&D Program in Japan. 14th Lighter-Than-Air.
6. Kuwano, Y. (1989): Progress of amorphous silicon solar cells. PVSEC4, Sydney, pp.14–17.

13

Future Prospects
for Photovoltaic Technologies in Japan

Nobuaki Mori and Toshihisa Masuda

The Japanese Photovoltaic Power Generation Technology Development Program (so called "Sunshine Program") was started immediately after the first oil crisis in 1973. Since that time, the meaning of PV technology has been changed from the generation of small-scale electrical energy to the development of bulk electrical power as an alternative to oil and to contribute to the suppression of global warming. As a result of having continuous R&D efforts on the technological development with market expansion aid by the government subsidies, Japanese PV industries have now started to work on self-development. In fact, the PV market in Japan has expanded markedly in recent years in the fields not only of the residential rooftop PV, but also other large-scale PV plants, such as public buildings and so on.

In this chapter, first the state-of-the-art on the photovoltaic technologies under the direction of the New Sunshine Project are overviewed, and recent achievements in the solar-cell production volume and the cost transitions are summarized. Then, future prospects of this new industrialization are discussed on the basis of a new strategy.

13.1 Current Status of Photovoltaic Industrialization

Figure 13.1 shows the transition of world photovoltaic (hereafter, called "PV") module shipments [1]. As can be seen in the figure, a marked growth of PV modules has been achieved, and reached global production of about 300 MW in 2000. Particularly rapid growth was seen in recent years in Japan, Europe, and USA. The production of PV cells has increased rapidly in Japan following the development of roof-type PV technologies and the introduction of a subsidy system to encourage the spread of PV systems.

This financial subsidy program for residential PV systems was started in 1994, and so Japanese production was initiated to increase rapidly since

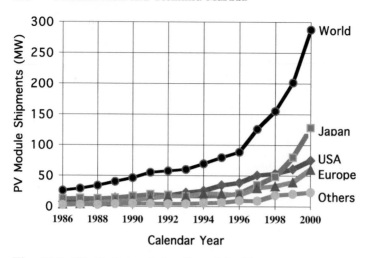

Fig. 13.1. World photovoltaic-cell module shipments

1994. Despite the recent recession in Japan, sales of PV systems are growing steadily, as the subsidy for roof-type PV systems has encouraged housing manufacturers to promote sales and install PV systems as standard on new houses. In addition, house buyers have been turning to PV housing in view of the lower environmental impact, regardless of the higher price.

The world solar-cell module productions of various types in recent several years are plotted in Fig. 13.2. The world total production in 2000 reached 288 MW and the rate of increase is 43% compared with that in 1999. The single-crystalline (c-Si) and polycrystalline (p-Si) Si solar cells are dominant in the production of photovoltaic (PV) cells. The production levels of c-Si and

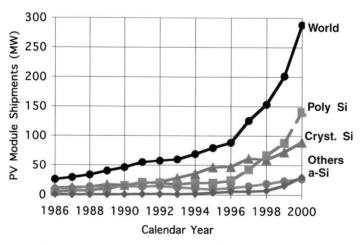

Fig. 13.2. World solar-cell module production of various types of solar cell

p-Si solar cells were 90 MW and 140 MW, respectively, in 2000. The growth rates of c-Si PV and p-Si PV are 23% and 60%, respectively. The increase of c-Si PV was large year after year and the production share was 49% in 2000. For other solar cells, the film-type a-Si PV production is almost constant and the production share was 9% in 2000. The ribbon-type PV production was 4.2 MW in 1999 and 14.7 MW in 2000 (growth rate of 250%), and had a production share of 5% in 2000. As a whole, crystalline-Si solar cells using bulk-Si materials still dominate in current PV market.

Figure 13.3 shows the cost transition of the various PV modules and components [2]. The price of PV modules decreases steadily with their mass production. These learning curves are similar to those of other electrical components with mass-production price changes. It is a very useful tool to predict the future module cost when a large introduction of PV systems is realized.

Figure 13.4 shows the calculated cost-reduction effect by mass production and the cost breakdown of a-Si module as one example. The cost-reduction rate is rather large until 100 MW/y production volume and not so large

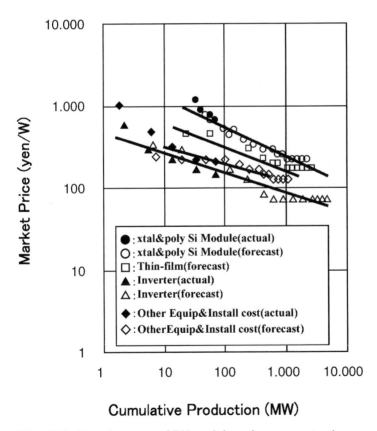

Fig. 13.3. Learning curve of PV-module and component price

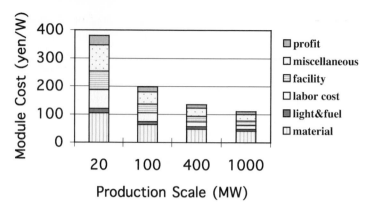

Fig. 13.4. Cost breakdown of a-Si module

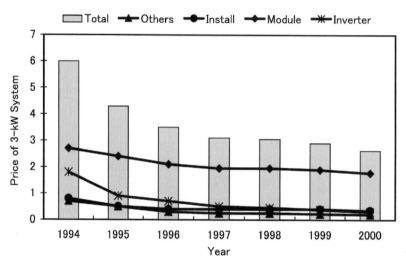

Fig. 13.5. Market price of residential 3-kW PV system (in mill Yen)

above 100 MW/y. The materials cost, labor cost, and facility cost account for the major parts of the module cost. It is very important to reduce the materials cost especially the glass base-plate cost and production-facility cost. In fact, changes in the market price of residential 3-kW PV systems are plotted in Fig. 13.5. As can be seen from this figure, the cost of peripheral devices such as inverters and control equipment, as well as installation cost has been decreasing, however, the cost of modules has not fallen markedly. PV R&D progress should directly help to reduce the module cost, but other factors such as production size or logistics cost appear to be having a strong influence.

The government subsidy for a PV house is reduced each year because of decreasing total installation cost by the increase of applications. The Residential PV System Monitoring Program was started in 1994 in order to disseminate

PV systems. In this program, the Government subsidized half of the installation costs. This program was continued until the FY1997 and changed to the Subsidy Program for Residential PV Systems that supports the installation costs based on the power output (kW). The subsidy for PV installation costs in FY1999 was 320 000 Yen/kW, and the number of applications was 18 000. The amount of the subsidy decreased down to 150 000 Yen/kW, and applications were 26 000 in 2000. Owing to this subsidy program, widespread introduction of PV systems reducing the installation cost in Japanese market has been promoted. Further reductions in PV costs are expected in FY2002.

One objective of R&D in phase 1 is to reduce the cost of manufacturing various PV modules to 140 Yen/W by the end of FY2000 (assuming a factory production level of 100 MW/y). These products will appear in the market around 2005. It is assumed that, at that time, the system sales price will be approximately 1 million Yen, and the PV power-generation cost will be estimated to be less than 30 Yen/kWh. As the market price of 3-kW systems is now about 2.7 million Yen, which is equivalent to a power-generation cost of 70–80 Yen/kWh, this target seems to be difficult to reach and it is necessary to take another approach to realize this target. Although this R&D objective can be accomplished technically, the market size has been expanding greatly and the breakeven point for PV-cell manufacturers is production of 100 MW or more per factory.

13.2 Recent Achievements of the PV R&D in the New Sunshine Project

After the first oil crisis in 1973, the Sunshine Program was initiated in 1974 to secure an alternative energy to oil and to ensure a stable supply of energy. Figure 13.6 illustrates the history of the PV National Program in Japan [3]. In the Sunshine Program, R&D on four new energy sources: solar energy (both thermal and photovoltaic), geothermal, coal-liquefaction and coal-gasification, and the utilization of hydrogen had been carried out. R&D on PV systems was also started in this program. Later, the new objective of preventing global warming was added to the technology development program, and the PV R&D program was rolled into the New Sunshine Program that was initiated in 1993.

The New Sunshine Program was originally a long-term project running until 2010, in which R&D was divided into two phases, phase 1 and phase 2. The R&D program of phase 1 has been completed by fiscal 2000. Although the progress of PV R&D, now in its 27th year since the first oil crisis, has not been fast, many results of the program have made it to market and performed well, showing that the program has been one of the more successful national projects. Figure 13.7 shows the chronological improvements in efficiencies of various solar cells in this program. An efficiency of 31.2% for the InGaP/InGaAs/Ge three-junction tandem cell of 25-cm^2 cell area has been achieved. The 14.6% for the thin-film mc-Si cell of 600 cm^2 area, 12.9% for

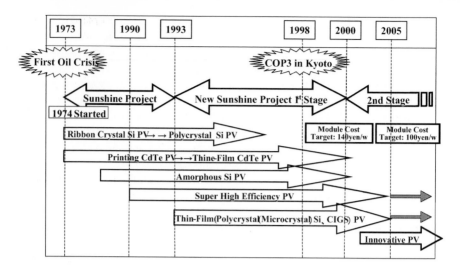

Fig. 13.6. History of PV National Program

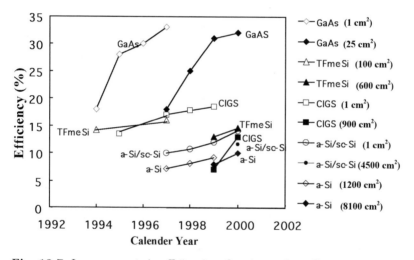

Fig. 13.7. Improvements in efficiencies of various solar cells

CIGS of 900 cm^2 area, 11.7% for the a-Si/poly-Si of 4500 cm^2 area, and 10% for the a-Si/a-SiGe of 8100 cm^2 area has been achieved.

Batteries play a very important role to improve the added value of PV power-generation systems. In recent years, a high-performance closed-type lead battery has been developed, which has a lifetime of ten years, and 3000 cycles of charging and discharging. Therefore, an inexpensive battery comparable to automobile batteries could hit the market if mass production is achieved.

Some solar cells and battery systems can generate uninterruptible power at prices comparable to normal electricity charge (around 24 Yen/kWh), combining with time-dependent charge systems (around 6 Yen/kWh at the late-night charge). As society becomes more information oriented and elderly, people will increasingly consider safety, health, and environmental protection, resulting in greater introduction of information technology into houses and complete usage of electrical devices. Uninterruptible systems with batteries will thus become essential for a stable supply of electricity. Furthermore, they will be required as an emergency power supply system in case of disaster. Also, when various energy-conservation technologies are developed, the market share of PV cells having small generation capacity but independent, distributed characteristics for portability and mobility will increase drastically, and so products utilizing PV power will be developed.

According to the preliminary assessment report of the Industrial Technology Advisory Committee that was issued in May 2000, the PV R&D project in the New Sunshine Project has proceeded well in many areas, such as improvement of PV cell conversion efficiency, development of implementing mass-production technology, grid-system development with public wires of utility companies, and the development of technologies for evaluating the performance of PV cells and systems. In order to further reduce the cost of PV cells and systems, the Industrial Technology Advisory Committee recommended starting phase 2 of the R&D program in fiscal 2001.

13.3 New Strategy and Future Prospects for PV Industry in Japan

Up to now, due to factors such as the prolonged low price of crude oil, renewable energies have faced difficult market conditions. However, recent awareness of the need for prevention of global warming is growing and the construction of new nuclear power plants, which have been plagued by accidents and troubles in Japan, has been delayed, so there are high expectations for renewable energies, especially PV power generation.

In March 2000, therefore, a tentative technology development strategy was reported at the sub-committee on strategy of new energy industrial technology development established within the Ministry of International Trade and Industry, aiming at strengthening the competitiveness of Japanese Industry related to new energy in the 21st century, creating new job opportunities and satisfying social needs such as preparing for an information technology and elderly society. Table 13.1 shows the new energy diffusion scenario toward 2030.

In the strategy, R&D of PV power generation was positioned as of particular importance, and a long-term technology development and promotion strategy until 2030 were proposed. Also, the government has recently organized the New Energy Working Group within the Energy Policy Advisory

Table 13.1. New Energy Diffusion Scenario toward 2030

New Energy Diffusion Scenario toward 2030

	1999FY		2010FY Forecast		2030FY Scenario		
	Installation	Oil equivalent	Installation	Oil equivalent	Installation	Oil equivalent	Ratio of Total primary energy
PV power generation	209 MW	-	4.8 GW	1.2 mil kl	53-82 GW	13-20 mil kl	1.7-2.7 %
Solar-heat utilization	-	0.98 mil kl	-	4.4 mil kl	-	7.5-10 mil kl	1.0-1.3 %
Wind-power generation	83 MW	-	3.0 GW	1.3 mil kl	2.1 GW	0.8 mil kl	0.1 %
Waste-power generation	980 MW	-	4.5 GW	5.9 mil kl	11-14 GW	18.8-28.5 mil kl	2.5-3.8 %
Waste-heat utilization	-	0.04 mil kl	-	0.14 mil kl	included in heat pump, etc.		
Biomass energy	-	4.6 mil kl	-	5.6 mil kl	-	15 mil kl	2.0 %
Heat pump, etc.	-	0.04 mil kl	-	0.58 mil kl	-	1.5-1.8 mil kl	0.2 %
Natural gas cogeneration	1.5 GW	-	4.6 GW	-	13-21 GW	-	-
Fuel cell (energy saving eff.)	12 MW	-	2.2 GW	-	17-31 GW	-	-
Total	3.5 GW	5.6 mil kl	17.1 GW	19.1 mil kl	93-150 GW	56-76 mil kl	7.5-10.1 %

Committee, and started discussing rules and an incentive mechanism for new energy technology development and promotion.

The Industry Technology Advisory Committee of the government has evaluated the phase 1 technology development program and decided to start the phase 2 technology development programs from fiscal 2001. To date, sales of roof-type PV systems have grown to create a large market, with subsidy support from the government. However, the market is not limited to house roofs. If the price of PV cells can be lowered, then PV systems could be used in various applications such as buildings, roofs of factories, and unused land. For instance, although this is not included in the current diffusion scenario, up to 80 GW of PV power could be gained by using 1600 km^2 of abandoned agricultural land that is no longer cultivated. Government is spending a lot of money every year keeping this land in a waiting circle for future reuse.

In Japan, sales of PV-generated electricity will clearly be larger than that of rice production per unit area of agricultural land. This idea could contribute to both food and energy security. To achieve this, continuous efforts to reduce cost are required, from the usual electricity charge (around 25 Yen/kWh), to the electricity charge for large consumers (10–15 Yen/kWh) or even to the power generation cost of utility companies (around 5–10 Yen/kWh).

Figure 13.8 shows the PV power-generation technology R&D roadmap that was formulated as part of New Energy Technology Strategies by the Ministry of International Trade and Industry, during the last half of fiscal year 1999. This PV roadmap provides the direction of PV R&D and promotion through to the fiscal year 2030. Advanced technology could achieve the target

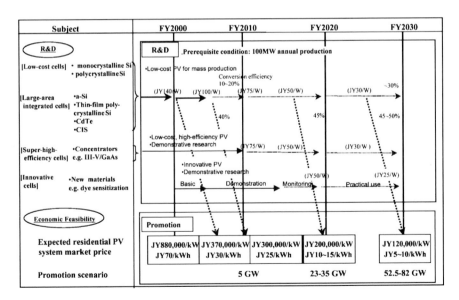

Fig. 13.8. PV power-generation technology R&D roadmap

of 25 Yen/kWh by fiscal year 2010. However, technological developments will not be sufficient for further breakthroughs. Indispensable will be innovative technology and innovative solar cells in order to attain the targets of 10–15 Yen/kWh by fiscal year 2020, and 5–10 Yen/kWh by fiscal year 2030.

Technological development for the reduction of the cost involves many issues such as cell cost, large-area integrated thin-film PV modules, next-generation new PV cells with R&D on low-cost recycling technology for the next age of mass production. Although the necessary cost level is very difficult to attain, Japan must set the target for the long-term R&D program aiming to bring the cost down to 5–10 Yen/kWh, because this is of prime importance to stimulate the PV market not only within Japan but also all over the world. R&D in the future will face even tougher situations, so further collaborative efforts will be needed to suppress global warming, stabilize the energy supply, and achieve sustainable development of society and the economy.

References

1. Maycock, P. (2001): PV News, PV Energy Systems, Inc.
2. Mori, N. (2000): 28th IEEE PVSC, Anchorage, Alaska, SP.0.1.
3. Mori, N. (2001): 12th Int. PVSEC, Jeju, Korea, S13.1.

Index

Printing: Saladruck Berlin
Binding: Stürtz AG, Würzburg